BUT WHAT IS THE RAINBOW?

EXPLORING THE SUPERVELOCITY AND SELECTION THEORY OF LIGHT

Intergrating Quantum Mechanical Superposition
with Velocity Space to attain a more
Parsimonious Version of Special Relativity

GEOFFREY STONE

But Where is the Rainbow?
Copyright © 2023 by Geoffrey Stone

All rights reserved. No part of this publication may be reproduced, distributed, or transmitted in any form or by any means, including photocopying, recording, or other electronic or mechanical methods, without the prior written permission of the author, except in the case of brief quotations embodied in critical reviews and certain other non-commercial uses permitted by copyright law.

tellwell

Tellwell Talent
www.tellwell.ca

ISBN
978-0-2288-9551-0 (Hardcover)
978-0-2288-9550-3 (Paperback)
978-0-2288-9552-7 (eBook)

Table of Contents

Introduction 1

Overview 1
   Physics or Philosophy? 1
   Thinking in Velocity Space 2
   Contrasting the Novel Theory with Special Relativity 2
   The Core Postulates of the Novel Theory 3
The Supervelocity and Selection theory of Light 5
   The First Clue is Found in the Rainbow 5
   Doubling Down on the Wave Theory of Light 5
   The Dynamics of Light 7
   A Photon is Born 8
   An Analogy between Position and Velocity 10
   Arbitrariness and Masking 13
   Refining and Testing the Theory 13
Precautions and Preparations 16
   The Challenge of Integrating and Embodying Knowledge 16
   Slaying the Hydra 17
   Compatibility and Conflict with Special Relativity 21
   Subjectivity, Objectivity and the Speed of Light 23
   Seven Questions to Reflect on 24

Relativity and Velocity Space 27

Is it Possible to See Light? 27
Leaving the Earth-Centred Inertial Frame of Reference 29
   Frames, Objects, and Observers 31
   The Galilean and Newtonian Revolution 33
   Defining your Location in Velocity Space 36
   Relative vs. Subjective and Absolute vs. Objective 38
Establishing an Objective Orientation 39

True Distance vs. Frame-Dependent Displacement 39
Frame-Dependent Displacement and Light 42
How Does Light Achieve Supervelocity? 47
Dynamically Defined Velocity: Light vs. Matter 58
Manifest Light vs. Latent Light 60
Integrating Three Critical Distinctions 63
Visualizing Velocity Space, Dark Matter, and Dark Energy 64
Relative Velocity and Simultaneity 74
Integrating Local Time with Global Time 80
The Need for Diversity in Both Measurement and Definition 83

The Finite and Infinite Manifestations of the Speed of Light 85

The Speed of Light as Infinite: Five Lines of Argument 86
Massless Photons Carry Momentum 86
Independence From the Source 87
Fixed Relationship to the Observer 88
Ageless Photons and Instantaneous Travel 88
Measuring the Distance to the Rainbow 89
Duality and the Speed of Light 90

Breaking the 299,792 km/s Speed Limit 91
Special Relativity and the Universal Speed Limit 91
Continuously Accelerating Spaceship 94
C as a Minimum Speed: C Only Applies to Light 96
Distant Galaxies Travel at Superluminal Speeds 96

Thoroughly Examining Special Relativity 115

The Three Corrective Distortions of Special Relativity 115
Einstein's Light Clock Thought Experiment 115
Time Dilation 117
Length Contraction 121
Inertial Mass Increase 122
Inertial Mass vs. Gravitational Mass 123
Common Themes in Special Relativity 124
The Hidden Complexity and Contradiction in Special Relativity 125

Does the Direction of Motion Matter? 125
   Thinking Outside of the Light Clocks 126
   Mapping the Subjective onto the Objective 128
   Where is the Rainbow? 129
   Light Speed Cannot Be Relative, Fixed, and Objective 132
   The Bidirectional and Unidirectional Speeds
   Must Be Different 135
   Einstein on Simultaneity and the
   Bidirectional Speed of Light 138
   The Time-Dilated Observer's Perspective 139
   The Two Opposing Types of Time Dilation 140
   The Manifestation of Time Dilation in Velocity Space 141
   The Three Distinct Levels of Inertial Time Dilation 142
The Twin Paradox and Symmetry 142
   The Preliminary Twin Paradox 142
   The Standard Twin Paradox 146
   The Twin Paradox 2.0 148
   The Nature of the Illusion of Inertial Time Dilation 150

Approaching the Speed of Light From Different Angles 151

   Superwave Dynamics 151
      The Pentality of Light 151
      Frame-Dependent Direction of Motion 155
   Applying the Analogy of the Visual Angle 157
      Visual Angle, the Moon, and Symmetry 157
      Visual Angle, Earth, and Infinity 162
      Visual Angle and the Three Speeds of Light 164
      Special Relativity as Obtuse Optics 168

Empirical Evidence and Predictions 173

   Falsifiability and Risky Predictions 173
   Empirical Evidence for Time Dilation and Length Contraction ... 175
      Magnetism and Length Contraction 175
      Muons and Time Dilation 177

Particle Accelerators and Time Dilation 179
Atomic Clocks and Time Dilation 180
Common Themes in the Purported Evidence 184
Predictions of the Supervelocity and Selection Theory 187
    The Basics 187
    Time Intervals depend on Locations in Velocity Space 188
    Intergalactic Oscillations in Time and Pulsars 189
    Rapidly Moving Mirrors 193
    Interpreting Red Shift with SSL Theory 195
    Conservation of Energy and SSL Theory 196

Important Questions 197

What is the Standard Deviation of Manifest Light? 197
    Manifest Light Must Have a Gaussian Distribution 197
    Absolute Precision by Definition 197
    How Thick is the Membrane of the Light Bubble? 199
What Is the Size and Shape of the Supervelocity? 201
Can we Shed Light on the Enigma of 137? 201
Is the Supervelocity Potential or Actual? 202
Further Questions about the Rainbow 203

Preview of Book Two 209

Understanding Entropy as Spherical Wave Dynamics 209
Probability as Subjective, Abstract, and Dynamic 210
Information is Created and Destroyed but Not Conserved 210
Conflating Information: Shannon, Laplacian, and Quantum 211
Characterizing Decision-Making Within Compatibilism 211
Retro-Causality and Decision-Making 212
Integrating Chaos Theory with Quantum Mechanics 212
Replacing Information Theory with Signal Detection Theory 212
On the Properties Ascribed to Elementary Particles 213
Quantization, Thresholds, Arbitrariness, and Randomness 213
The Inherent Inversion of Quantum
Mechanical Representations 214

Glossary 215

Bibliography 221

Introduction

Overview

Physics or Philosophy?

I could place this book under the category of philosophy due to my philosophical approach to the subject matter as well as my academic background and interest in philosophy. In these pages you will find a heavy reliance on thought experiment and analogy. I am adhering to an overarching argument structure that is coherent, organized, complex, and broad in scope. Much of my focus is on defining, refining, and clarifying difficult concepts and the questions that surface significantly outnumber the conclusions that are reached. Overall, this book has a distinctly philosophical flavour.

On the other hand, the subject matter of this book is centred in theoretical physics. In particular, I deal with the speed of light, special relativity, and my own Supervelocity and Selection theory of Light. This work also ventures into the neighbouring territory of metaphysics, psychophysics, and the philosophy of science.

I had nearly finished this lengthy and all-consuming project before I finally stumbled upon the insights that led me to the Supervelocity and Selection theory of Light. This was an unexpected and fortunate revelation. The puzzle pieces that I had been studying had all seemed to come together at the very last minute, and ever since that moment, the revised goal of this project has been that of introducing the world, and the physics community, to this new physical theory. Whether this book should be considered physical or philosophical is of relatively minor significance.

Thinking in Velocity Space

In the same way that both special and general relativity rely on a model known as "spacetime," the Supervelocity and Selection theory of Light relies on a model which I call "velocity space." Just as one must learn to think in terms of spacetime in order to grasp special relativity, one must also learn how to think in terms of velocity space in order to grasp the Supervelocity and Selection theory of Light. Fortunately, as both a mathematical and conceptual model, velocity space is far simpler, and easier to grasp, than is spacetime. Velocity space is also far more intuitive and far easier to work with, as compared with spacetime. Velocity space is just a three dimensional space where all points represent velocities and all distances between points represent relative velocities. Unfortunately, if the reader has grown accustomed to thinking in terms of spacetime, as many physicists and engineers have, then their background, and experience, may hinder their ability to think in terms of velocity space. Therefore, making the conceptual transition towards thinking in terms of velocity space may require some additional time and effort for those who regularly work with special or general relativity.

I have always been inclined to analyze and interpret the subject matter of special relativity in terms of velocity space. However, it took a long time before I was able to explicitly recognize that I was doing this. This unique method has always struck me as the most obvious and logical approach to take. After dozens of in-depth conversations with various physicists, I was eventually able to appreciate how rare it is to approach this topic through the lens of this fruitful and indispensable model.

Contrasting the Novel Theory with Special Relativity

According to special relativity, time and space must both bend in order to allow observers who occupy different locations in velocity space to see the same light move at the same relative speed. According to the Supervelocity and Selection theory of Light, time and space are not required to bend in order to reconcile these otherwise divergent observations because the light that is seen by each observer is in some sense not really the same light. According to the Supervelocity and Selection theory of Light, observers who occupy different locations in velocity space will see different light because the light that is seen has been selected independently by each observer. This is the difference between special relativity and the Supervelocity and Selection theory of Light in a nutshell.

The main advantage that the Supervelocity and Selection theory of Light has over special relativity is that it explains the observer-centred and observer-dependent nature of the speed of light much more efficiently and effectively than does special relativity.

The Core Postulates of the Novel Theory

I can express the essential thesis of the Supervelocity and Selection theory of Light in a single sentence, as follows: Light does not travel at a particular speed, but light does travel at a range of speeds.

The "range of speeds" is now referred to as a *supervelocity*, and the particular speed that is eventually observed is only the end result of a selection process which the observer unwittingly engages in.

I understand why it would strain credulity to be presented with the claim that light doesn't travel at any particular speed given that it is universally acknowledged that the speed of light is exactly 299,792,458 m/s in a vacuum. At first blush, it looks like I am contradicting one of the most important and well established facts in physics. However, these two seemingly contradictory claims are much more compatible than they appear to be. The speed of light is recognized as a special speed, and deservedly so. As you turn the pages, it will become increasingly clear that there is an astonishing degree of depth to consider in relation to the nature of the speed of light.

The core postulate of the Supervelocity and Selection theory of Light is as follows:

1. **An object has a defined velocity if and only if it has a defined location in velocity space.**

    From this postulate it necessarily follows that light does not have a defined velocity because the location of light cannot be defined in velocity space. 299,792,458 m/s relative to all observers is not a defined location in velocity space. By contrast, it is possible to define the locations of all material objects (such as planets, stars, people, and cars) in velocity space.

    The Supervelocity and Selection theory of Light also contains the following three additional postulates.

2. **Latent light occupies a superposition in velocity space.**

3. **Each observer independently selects manifest light from the available latent light by virtue of that observer's objective location in velocity space.**

4. **All of and only the latent light that resides in velocity space at a distance of c from the observer is selected.**

In addition to these four postulates, the Supervelocity and Selection theory of Light is also founded on the two core postulates of special relativity. The two core postulates of special relativity are based on the results of the Michelson-Morley experiment in 1887 which determined that the relative speed of light does not change as Earth orbits the Sun or as the direction of a beam of light changes. The two core postulates of special relativity are as follows:

1. **The laws of physics are the same in all inertial frames of reference.**

2. **The observed relative speed of light is the same in all inertial frames of reference, irrespective of the speed of the light source relative to the observer.**

The Supervelocity and Selection theory of Light

In the sections that follow I will illustrate a broad sample of my ideas with a short narrative (The Dynamics of Light/A Photon is Born) which provides the essential details of a novel theory that really does have the potential to unite quantum mechanics with relativity and to topple the existing paradigm. This narrative will be extremely dense and fast-paced in comparison to the rest of this book, so please do not be discouraged if you find it hard to digest. Before diving into the narrative, I will provide some more context.

The First Clue is Found in the Rainbow

The arc of a rainbow appears to stand in a particular position, but it really doesn't. By the same token, the spherical wave front of a flash of light appears to expand outwards in all directions at a particular speed, but it really doesn't. This is the core insight behind the Supervelocity and Selection theory of Light, and I invite you to pause and reflect on it before I develop it further.

Doubling Down on the Wave Theory of Light

Until the twentieth century, the various particle and wave theories of light were in direct competition, and both had accrued comparable success and recognition. In the twentieth century, the prominent theories of light had become those which maintained either a moderate compromise or a balanced integration between particle models and wave models. "Wave-particle duality" is the generic description for this modern hybrid class of theories. The Supervelocity and Selection theory of Light (SSL theory) is *not* a hybrid wave-particle duality theory, as it solves the problem of how light can manifest as both wave and particle in a radically different manner.

According to the Supervelocity and Selection theory of Light, light is neither a particle, nor a wave, but a superwave, as seen from a specific vantage point. Light is even more wavelike than how it has been previously conceived. Light is more wavelike than either surface waves or sound waves. By comparison to a superwave, an ordinary wave is relatively particle-like. The manifestations of the superwave are more aptly described as a *pentality* than as a duality. As I will explain in due course, it is the fact that light is fundamentally a superwave, which simultaneously explains both its wavelike characteristics, and its particle-like characteristics.

I want to point out that while *supervelocity* is a novel term it is not an entirely novel concept in quantum mechanics, and therefore the idea of a supervelocity should not inspire shock or disbelief. A supervelocity more or less corresponds to what would be referred to as a *quantum superposition of momentum eigenstates,* in the language of quantum mechanics. Energy and momentum are closely related to velocity, and momentum is included as a variable in the Heisenberg uncertainty principle. Hence, it is already widely acknowledged in quantum mechanics that elementary particles can and do have vaguely defined velocities and that a particle may occupy a range of different velocities at the same time. A superposition in the property of a particle may manifest as a distribution of potential positions, velocities, speeds, directions, spin directions, or moments in time. A particle that is in quantum superposition will only have a range of potential observed values which are described by probability distribution. I am merely borrowing this concept of quantum superposition and applying it to the velocity of electromagnetic radiation in general, while also hypothesising a non-random selection mechanism with which the observer is able to collapse the probability distribution. But I am jumping ahead too far... let's start at the beginning!

The Dynamics of Light

Let's begin with an explanation for the observed velocity of light. Imagine that a tiny flash of light is spontaneously emitted in a relatively dark and empty region of space. The location of the origin of this instantaneous flash of light has been precisely determined, along with the moment in time when it occurred, but there is a lot that still remains to be determined. What should the velocity of this light be?

This pulse of electromagnetic energy then proceeds to emanate from its original location at all velocities simultaneously. Because velocity is a concept which incorporates both a direction and a speed, this means that the pulse is travelling in all directions simultaneously, and at all speeds simultaneously.

The reason that this pulse of energy *must* travel at all velocities simultaneously is that any choice of a single velocity would be arbitrary, and so there is no way to set any rule or guideline in advance in order to constrain it. Because no choice can be made in a principled manner, no choice is made. If you are wondering why, the velocity of this light takes on all possible values simultaneously, consider the fact that according to Galilean relativity, special relativity, and general relativity, all inertial frames of reference are equivalent. In other words, there is no preferred frame of reference that one could use in order to define the velocity of any object in absolute terms. At first the light is totally free because it is not tethered in any way to any particular object or to the frame of reference of any particular observer.

But now let's add an observer to this scenario. In addition to being an observer, this observer is also an object. This observer also has no particular absolute velocity, but the observer does have an infinitely large set of relative velocities which are each relative to various other objects. Those other objects also have their own sets of relative velocities with respect to each other and with respect to this observer. This network of relative velocities contains invariant properties which can be used to objectively define the inertial frame of reference of this observer. The observer and all of the objects that appear to be motionless relative to the observer would all share a common objective location in velocity space.

One fascinating but subtle feat that this observer's objective location in velocity space accomplishes is that it limits what an observer is able to interact with and therefore also what they are able to observe. This observer can only interact with and subsequently observe electromagnetic energy if it is moving relative to the observer at a speed of roughly 299,792 km/s. As a necessary consequence of this constraint on the observed speed, what you end up with, from the point of view of the observer, is what looks like an expanding sphere of energy which grows at a rate of roughly 299,792 km/s in all directions at the same time. What this means is that the speed of the pulse has now been determined, but only for this particular observer, and only relative to this particular observer's location in velocity space. In a way, the overall behaviour and geometry of this pulse of energy is beginning to resemble that of a sound wave, but the precise direction of this pulse of energy remains to be determined.

A Photon is Born

Now, let's add a new detail to this story to illustrate a few quantum mechanical concepts. This flash of light still has not been observed. This flash of energy does *not* travel through any medium and it is *not* an emergent phenomenon built from a cascade of smaller causal interactions between discrete particles. Despite the fact that it lacks a few of the fundamental characteristics of all known waves, we still attribute some wavelike properties to this pulse of energy. We say that this energy has a particular frequency and also a particular amplitude. In this case, the frequency of this flash of light is high enough and the amplitude is low enough such that one and only one discrete quanta (or photon) of energy is emitted. Before this pulse of energy interacts with anything it really is travelling in all directions at the same time. Because this photon is moving in all directions at the same time, we say that it is in *superposition*.

Finally, this photon interacts with an observer. At that moment, its velocity becomes fully defined in terms of both its speed and its direction. We can now determine the precise location where this quanta of energy interacted with a specific molecule, and we can also reconstruct its direction of motion under the assumption that it took the shortest available path from point A to point B. Because the location of the interaction that finally takes place is extremely focused and point-like, we say that the photon has a particle-like quality to it. However, the direction of the motion of this quanta of energy must also be arbitrary, as its direction is a component of its velocity, which we have already established as arbitrarily determined.

This singular quanta of energy could not have remained in a loosely defined state of superposition forever. If it is to interact with any particular object, then the directional component of its velocity must become narrowly defined. Because matter is made up of discrete molecules, because a one-to-one correspondence between cause and effect must be maintained, and because the potential for a causal interaction must be maintained indefinitely, this means that one molecule and only one molecule will interact directly with this tiny flash of energy.

Even once the speed component of the velocity of the flash has been determined, this still leaves the directional component unresolved. The absence of a principled way to determine the direction of the motion of this quanta of energy, combined with the need for it to be extremely narrowly defined, results in the seemingly random placement of the resulting interaction. This interaction is normally attributed to the impact of a single photon particle colliding with a single molecule. The particular molecule that will be affected is selected randomly from all available molecules and this is the logic behind what is known as *quantum randomness*. If the full superposition represents all of the light, and the tiny photon represents only a small fraction of the light, then it is pretty clear that nearly all of the light is never observed because it can never interact with anything except for itself.

There appear to be three stages and two transitions visible here. We begin with a superposition of all velocities, and then this superposition of velocities is narrowed down when one speed is selected, and it becomes a superposition of all directions. This superposition of all directions has many wavelike qualities to it and it is also narrowed down to a single direction through its interaction with an observer. This single direction is selected in a seemingly random manner, and it has a particle-like quality to it. In other words, a superwave becomes increasingly defined as it transitions into a wave, which then becomes increasingly defined as it transitions into a particle. This pair of transitions is somehow caused by the act of observation. Prior to observation, the only defined properties that the superwave possesses are its amplitude, frequency, point of origin, and time of origin.

An Analogy between Position and Velocity

Just to emphasize the striking parallels in logic, I have included two matching paragraphs below which give analogous descriptions of two related phenomena. These paragraphs compare and contrast the relationships between the observer and each of the two components of light's velocity vector. The point of this exercise is to bridge the gap in understanding from one set of claims which are easy to accept, to an analogous set of claims which are relatively challenging.

The direction component:

When observing a bright but distant star, an observer who is located in a particular position will only be able to see the light that travelled directly from the point of origin to their specific location. That is, they can only see the light that travelled in one, very specific direction, while the rest of the light is not observed. Meanwhile, two different observers who are each located in two different locations will see the same light move in two different directions. They will each see the portion of the light that moved in a direction that is specifically tailored to their particular location in space. In a relatively superficial sense, both observers will see the light moving in exactly the "same" direction, because according to both observers the light moved "directly towards me." An important part of what it means to be in a particular location is to only be able to see a very small proportion of the total light and only that light which moved in a particular direction such that it was able to pass through the narrow opening of your pupil. Multiple observers in different locations can also integrate their seemingly contradictory observations in order to confirm that the light must have moved in multiple directions at the same time. The direction of the motion of the observed light was actually determined by the relative position of the observer.

There is nothing shocking or challenging about the above paragraph, yet I anticipate that the following paragraph will raise a few eyebrows.

The speed component:

When observing a bright but distant star, an observer who is coasting in a particular inertial frame of reference will only be able to see the light that is travelling at roughly 299,792 km/s relative to their inertial frame of reference. That is, they can only see the light that travelled at one very specific speed, while the rest of the light is not observed. Meanwhile, two different observers, who are each moving in different inertial frames of reference, will see the same light move at two different velocities. They will each see the portion of the light that moved at a velocity that is specifically tailored to their particular inertial frame of reference. In a relatively superficial sense, both observers will see the light moving at exactly the "same" speed, because according to both observers, the light moved "at 299,792 km/s relative to me." An important part of what it means to be in a particular inertial frame of reference is to only be able to see a very small proportion of the total light and only that light which moved at a particular speed, such that it was able to interact with the light-sensitive cells in your retina. Multiple observers in different inertial frames of reference can also integrate their seemingly contradictory observations, in order to confirm that the light must have moved at multiple speeds, at the same time. The speed of the motion of the observed light was actually determined by the relative velocity of the observer.

Comparing the two paragraphs side by side, and sentence by sentence, really makes the many parallels more apparent.

There are two reasons that the second paragraph is more difficult to accept than the first. One reason is that the basic principles of optics are highly intuitive, and we have a lot of daily experience with the directional component of the velocity of light. By contrast, we don't have any daily experience with the speed of light. We didn't even know what the speed of light was 400 years ago, nor did we know whether this speed was finite or infinite. The second reason is that this second paragraph blatantly contradicts how theoretical physicists currently think about the speed of light and how they teach the fundamentals of special relativity. The current received wisdom is that light moves at a singular, finite, and constant speed, and *not* at a supervelocity comprising multiple speeds and directions.

Arbitrariness and Masking

Arbitrariness is one way of understanding what is behind the various types of indeterminacy that we see in both special relativity and quantum mechanics. It is also possible to understand the indeterminacy as a means of hiding or masking knowledge. Seeing a constant speed of light can be construed as way of making it impossible to determine any absolute frame of reference. Arbitrariness can also be viewed as the initial cause of indeterminacy. It may be that the masking or hiding of knowledge is the purpose of this indeterminacy and of the arbitrariness by extension. In quantum mechanics this arbitrariness manifests as random patterns, whereas, with the speed of light, this arbitrariness manifests as predictable patterns which are anchored to the observer's location in velocity space.

Refining and Testing the Theory

While the Supervelocity and Selection theory of Light (SSL theory) does tie up a number of loose ends, it is still very much an underdog when compared to the established track record of special relativity as currently formulated. The most direct way to refute the Supervelocity and Selection theory of Light is to point to the empirical evidence that has been amassed to support both inertial time dilation and length contraction. I will thoroughly address the purported evidence for inertial time dilation and length contraction towards the end of this book.

The idea that light travels at all relative speeds simultaneously until it is finally observed is a new idea. According to the Supervelocity and Selection theory of Light, the fact that light sometimes appears to travel at a single, finite speed is a consequence of a somewhat mysterious selection mechanism. This selection mechanism might be analogous to the way that a transistor radio filters out all frequencies of electromagnetic radiation except for the narrow band of frequencies that the dial is set to receive. The selection mechanism that I have in mind filters out all relative speeds except for the relative speed that we recognize as the speed of light and this process is somehow guided and calibrated by the location of the observer in velocity space. This selection process is determined by a deeper logic of interaction that should be investigated in more depth.

While the selection mechanism in the Supervelocity and Selection theory of Light is mysterious, it is not nearly as mysterious as the comparable selection mechanisms that operate in quantum mechanics. As with these quantum mechanical selection mechanisms, we don't yet know how to inspect this selection mechanism more closely in order to see the details of how it works. But unlike the quantum mechanical selection mechanisms, we can predict in advance exactly which choice it will make. So, we do already have a superior understanding of this selection mechanism as compared to the corresponding quantum mechanical selection mechanisms.

One way to confirm and apply the Supervelocity and Selection theory of Light (SSL theory) would be to discover new ways of manipulating light so that latent light, which is not ordinarily visible, can be made visible. Just as carefully positioned mirrors and lenses can be used to alter the directional component of the velocity of light, perhaps mirrors, and lenses with different velocities can also alter the speed component of the velocity of light. We can also study distant pulsars and type I Cepheid variable stars in order to find the predicted temporal oscillations.

The Supervelocity and Selection theory of Light explains some of the wavelike behaviors of light. It explains why light normally appears to expand outwards in all directions simultaneously in the form of a growing sphere. The apparent reason for this being that this is precisely the pattern that you are left with if you start out with all possible velocities and then you eliminate all of the velocities that don't meet the narrow selection criterion of 299,792 km/s, relative to the observer.

Precisely how narrow this selection criterion is, has not yet been determined. It would likely have an average value that is close to 299,792 km/s, but also a Gaussian distribution of a certain size surrounding that average value. The observed width of the Gaussian distribution would depend on the assumptions and the methods of the measurement process. If experiments are performed very carefully, then we should be able to observe a residual Gaussian distribution, which would imply that the relative speed of manifest light is actually a narrow, and normally distributed range of speeds, with an average value that corresponds to the recognized value of **c**.

According to the Supervelocity and Selection theory of Light, the frequency of vibration of light remains finite and fixed. Meanwhile, both the speed and the wavelength of the light are in superposition. Faster speeds correspond to longer wavelengths while slower speeds correspond to shorter wavelengths and the standard equation for this is: speed = wavelength x frequency. The relationship between speed and wavelength can also be observed in the refraction of light as it travels through various media. When light traverses a medium it will reduce its speed and also shorten its wavelength to one degree or another depending on the refractive index of the medium.

Another reason for why I expect to see a lot of resistance to the Supervelocity and Selection theory of Light is due to a few very well-established assumptions about the nature of light. Light has historically been viewed either as a wave or as a particle. It is clear that sound waves and surface waves do not travel at multiple speeds simultaneously. Large particles also do not travel at a wide range of speeds simultaneously. Therefore, how could light travel at all speeds simultaneously?

Because the Supervelocity and Selection theory of Light (SSL theory) posits that light is neither a wave, nor a particle, but a superwave, its speed is not constrained in the way that that the speeds of particles and waves are. Light is to a wave, as a wave is to a particle. Unlike particles, waves are capable of passing through each other without interacting and also moving in all available directions at once. Unlike waves, superwaves occupy all velocities at once, they occupy all polarities at once, and they have an amplitude which has no spatial extension. A superwave doesn't travel through what we would ordinarily recognize as a medium, although it can also be modeled as a wave that travels through a very special medium which occupies a superposition in velocity space. Like empty space, the medium of a superwave would contain no discrete moving parts. The resonance and interaction of superwaves also seems to operate according to a different set principles from that of ordinary waves.

Precautions and Preparations

The Challenge of Integrating and Embodying Knowledge

As we are exploring the deeper mysteries of the universe, it becomes clear early on that we are lost. The human mind was not designed to allow us to wander this far from our natural environment. Therefore, it is not enough to merely know something if it happens to be something that is deeply unfamiliar to us. We may need to be reminded regularly of facts which are not congruent with our other beliefs or with our habits. We often need to rigorously retrain our brain in order to apply new knowledge. We also need to take the time to integrate new knowledge with our other beliefs and with our behavior, in order to successfully demonstrate that we truly understand it, know it, or believe it.

Remaining calm in the presence of a garter snake demonstrates your knowledge that the snake is harmless. Refraining from smoking demonstrates your knowledge that smoking is harmful to you. Watching a movie without responding emotionally to it demonstrates your knowledge that it is just a movie. It is the full integration and application of our beliefs and principles which should be the standard by which we judge whether or not we actually hold them. A mere endorsement of an idea is not sufficient. In other words, if knowledge does not fully possess us then we do not fully possess knowledge.

Slaying the Hydra

It should not be surprising that we have not yet identified any particular universal limit on the human understanding of reality because to precisely identify any such limit would also imply some degree of transcendence. If such a limit does exist, then it can't also be found, for as soon as it has been identified, then it has already ceased to exist. And yet, it is easy to find phenomena that are either challenging or seemingly impossible to understand at present. The persistence of this confusion may be explained by pointing to its temporal and dynamical aspects as well as its collective and interactive aspects. The whole morass of confusion is much greater than the sum of its parts, and this whole is also unique to each individual, and to each point in time. The bits of confusion can be conquered relatively easily in isolation, but the whole complex is at the same time dynamic, persistent, and elusive.

As I have mentioned above, stating the correct answer to a question is the easy part. However, fully, and consistently integrating and applying the corresponding knowledge is relatively hard. This stubborn challenge brings me to the Hydra of confusion that is impeding our progress in grasping the deepest mysteries of reality.

In Greek Mythology, the Hydra is a giant beast with multiple heads mounted on long necks which are attached to the same body. To slay a Hydra, one would naturally attempt to cut off its heads one by one. But this strategy will not succeed, because so long as the beast has a few heads remaining it will still be alive. So long as the Hydra is alive it also has the unique ability to rapidly heal and regenerate any heads that have been cut off. Therefore, while it may appear as though you are making progress in slaying the Hydra, in reality you are getting nowhere.

Perhaps you have found yourself in an argument with someone and noticed that the discussion was going in circles. What had been established just two minutes ago, has already gone down the memory hole, and into oblivion. Slaying a Hydra is more challenging than it appears because the Hydra maintains the insidious capacity to regenerate. The adaptive and dynamic nature of the beast ensures that what appears to work very well over the short term will ultimately fail. The Hydra that I have in mind is almost impossible to slay without a combination of persistence and the implementation of a masterfully executed strategy.

What is needed is patience, practice, and repetition. The heads of the Hydra also need to be rendered robustly recognizable. These heads need to be seen from a wide range of angles and in a variety of contexts. One must be able to recognize this beast even while it is in camouflage and nestled within its natural environment. Standing in our way will be our own natural tendency towards hubris and complacency. You may wonder how it is possible to unconsciously assume that a particular idea is true in one context if we are also able to easily recognize it as false in another context. This actually occurs quite readily, easily, and frequently. As a general rule, our thought processes are mainly automatic, and our beliefs are anything but consistent and rational.

As a consequence of the characteristics of the beast that we are attempting to slay, some of the positions that I oppose will appear convincingly as straw-man positions, even though they are not. Many of the positions that I will be taking and the claims that I will be making will also seem blindingly obvious, but this is only because I have carefully situated these ideas in a context that renders their validity obvious.

The most sophisticated engineers, physicists, and mathematicians are also human. This special breed of experts is no more immune to the influence of misconceptions, false assumptions, and fashion trends than the rest of the species is. Nobody ever said that unravelling the deepest mysteries of reality would be easy, even if that appears to be your primary occupation.

I have constructed an enumerated list of the seven heads on the Hydra of confusion:

1. Confusion between the objective location of an observer in velocity space and the arbitrary location of the zero in velocity space. The common phrase *inertial frame of reference* is normally used to refer to both of these concepts and it also ties them together. Subjective framing is therefore tied to an objective location in velocity space. In this way, the *relative vs. absolute* dichotomy is conflated with the *subjective vs. objective* dichotomy. This confusion leads many to the false conclusion that all velocities are necessarily subjective, merely because they are relative. In order to avoid confusion, these two dichotomies must be treated as separate and distinct. The use of velocity space will assist us in distinguishing between these dichotomies and in making it clear that velocities can be both relative *and* objective.

2. Confusion between the objective true distance between two objects and the frame-dependent displacement of a moving object. Unlike the frame-dependent displacement, the true distance does not depend on an arbitrary frame of reference. The length of the path taken by a beam of light is a frame-dependent displacement, as opposed to a true distance. Because the frame-dependent displacement of light is regarded as fixed, it is mistakenly conceived of as objective. Conflation between these two concepts makes the true distance appear to be just as subjective as the

frame-dependent displacement is. This confusion is also fed by the assumption that distance can only be derived from multiplying velocity by time. Further confusion is caused by the assumption that distance rests on a subjective assessment of simultaneity, which in turn depends on a relative velocity. It is important to keep in mind that there are other ways of both measuring and defining distances which are relatively stable and objective.

3. Confusion between manifest light and latent light. Because the existence of latent light has not been previously acknowledged, it was therefore assumed that the manifest light that one observer sees from one location in velocity space must also be the same manifest light that another observers sees from another location in velocity space. The reality is that the particular latent light which manifests for two different observers will differ depending on where each of them is in velocity space. The two observers are not both viewing the same light and so there is no real discrepancy to reconcile.

4. The idea that the speed of light can be simultaneously characterised as fixed, relative to the observer, and objective. If any finite speed is both fixed and relative to the observer, then it must be subjective. For any finite speed to be objective, then it cannot be both relative to the observer and fixed. In chapter four I will use an analogy with a rainbow to make this point more clear.

5. The idea that objects or observers are able to move through space itself, or relative to space itself, along with the idea that space itself can move, stretch, expand, or shrink. This is a subtle resurrection of the idea of an absolute inertial frame of reference, and it contradicts the foundations of special relativity.

6. The idea that the relative and finite speed of 299,792 km/s also functions as a speed limit. This special speed is not only considered to be the speed of light but also the speed of gravity waves, causality, and information. This would have been plausible if 299,792 km/s had been either defined in velocity space or infinite, but because it is neither defined in velocity space nor infinite, then it cannot function as a speed limit.

7. The idea that symmetry is an epistemological limit. The question of who is "really" in inertial motion or who is "really" stationary is frequently taken as a meaningful question in this context. The continuous dimension of acceleration is mistakenly applied as a means of deciding the winner in a contest between two incompatible accounts of reality. This is also a subtle resurrection of the idea of an absolute inertial frame of reference. In chapter four, and in the context of the twin paradox, I use the analogy of a poker game to expose the absurdity of this conception of symmetry.

Compatibility and Conflict with Special Relativity

I don't want to create the impression that I am challenging or dissenting from special relativity per se because that would not be a fair characterization of my position. The Supervelocity and Selection theory of Light (SSL theory) is much more compatible with special relativity than it may seem to be at first glance. Special relativity is also a complex theoretical structure which developed gradually over time and contains multiple layers and facets. If the Supervelocity and Selection theory of Light is eventually accepted as valid, then this only implies that the more superficial features of special relativity have become redundant.

I fully accept the core foundations of special relativity which are based on the results of the Michelson-Morley experiment, and which include the two postulates of special relativity. Just as with special relativity, the Supervelocity and Selection theory of Light is also founded on the idea that light is always observed to move at **c** relative to each observer, and that the velocity of a light source has no influence over the observed velocity of the light that it emits.

As Einstein and others developed special relativity, they have made a number of implicit assumptions which would have seemed necessary but were actually optional, and it is at the point of making these optional assumptions that the Supervelocity and Selection theory of Light departs from special relativity. The Supervelocity and Selection theory of Light (SSL theory) does not rely on Einstein's light-based definition of simultaneity, on his reasons for proposing inertial time dilation and length contraction, on the idea that the speed of light is a universal speed limit, or on the Minkowski spacetime manifold.

What I have developed is an alternative account of the foundations of special relativity which is derived from the same empirical evidence. My method for integrating the divergent perspectives of observers who occupy different locations in velocity space is radically different from that of Einstein's. My method is both less complicated and more comprehensive than Einstein's method. Instead of spacetime, the Supervelocity and Selection theory of Light (SSL theory) relies on an objective model of *velocity space*. Instead of inertial time dilation and length contraction, my theory relies on the objective conceptions of *true distance* and *latent light,* where true distance is the distance between two objects without any consideration of either velocity or time, and latent light is all of the light that exists, including the light which cannot be seen from a given location in velocity space.

What I am really doing with SSL theory is reaffirming the roots of special relativity while extending it in a new direction that is informed by quantum mechanical insights which were not available in 1905, when Einstein first proposed his theory.

Subjectivity, Objectivity and the Speed of Light

Because I will be relying on the critical distinction between objective and subjective very heavily, it is important that I am not misunderstood. When I use the words *objective* and *subjective* to represent opposing extremes on a spectrum of perception, I have a very precise and nuanced definition of these terms in mind. According to the definition that I will be using, as we move from subjective to objective perception, we are doing the following two important things simultaneously:

1. Resolving ambiguities

2. Correcting biases.

To advance in the direction of objectivity we need to construct a new model of reality which is flexible and comprehensive enough to consider and integrate a wider range of perspectives.

To borrow an example used by Thomas Nagel in his book, *The View from Nowhere*, the distinction between subjective and objective is the distinction between looking at the world with monocular vision and with binocular vision. The difference between these two methods of perception is subtle, but the nature of this difference is clear. Binocular depth perception is an excellent concrete example of relative objectivity.

Another good example of increasing objectivity is to think about how a cylinder may appear as a circle from one angle and as a rectangle from another angle. Both the square and the circle hypotheses which may be entertained are relatively subjective, but the cylinder hypothesis is relatively objective. Plato's allegory of the cave is also a helpful metaphor to describe the process of obtaining objective knowledge. This distinction between subjective and objective will be one of the most important and frequently used distinctions in this book, and so it is very important to be on the same page, semantically speaking.

A closely related conceptual distinction that I will be relying heavily on is that of *latent vs. manifest*. It is helpful to think of this distinction with an iceberg analogy. We can see only the tip of the iceberg, but we also know that most of the iceberg is hidden beneath the surface. Along the same lines, if someone has a latent illness, this illness will manifest in symptoms that you can see, but you cannot see the latent illness directly.

Seven Questions to Reflect on

Before I close this introductory section, I want to conclude with seven revealing hints about light for the reader to reflect on:

1. Whenever you hear it correctly stated that it is impossible to catch up with a light beam, remember that it is also impossible to touch a real rainbow. Why might that be?

2. Whenever you hear it correctly stated that it is impossible to travel fast enough to escape from a light beam, remember that it is also impossible to run fast enough to escape from your own shadow. Why might that be?

3. Whenever you hear it incorrectly stated that nothing can travel faster than 299,792 km/s, remember that it is always possible to accelerate at any rate and in any direction that you want to, without limits. Why might that be?

4. Whenever you hear it misleadingly stated that the speed of light is a "constant," remember that this same property also applies to the direction of light because for light to be observed it must always be pointed directly towards the observer. Why might that be?

5. Whenever you hear it misleadingly stated that the speed of light is fixed, remember that so is the apparent distance between an observer and a rainbow. Why might that be?

6. Whenever you hear it stated that each observer thinks that the other observer's time is dilated, remember that each observer also "thinks" that the other observer appears to have shrunk, because more distant objects always look smaller. Why might that be?

7. Whenever you hear it stated that the maximum speed is the finite value of 299,792 km/s, remember that the maximum visual angle is also the finite value of 180 degrees. Spherical objects cannot appear to be any larger or closer than 180 degrees. Why might that be?

Relativity and Velocity Space

Is it Possible to See Light?

We don't really see light per se. Instead, we see various objects which make their presence known through the way that they affect the dynamics of the ambient light. Light helps us to see objects but there is really no need to see light itself. Even if you stare directly at a source of light, such as the Moon at night, you are not really looking at the moonlight. Rather, you are looking at the Moon. By contrast, the Moon's light is essentially everywhere, and it is a good thing that you can't see all of it. One way of understanding why it is that you don't really see the moonlight is to think about how at least 99.9999999 percent of the sunlight that reflects off the Moon will never enter either of your pupils. Light must enter a pupil in order to be seen, and yet a pupil is a very small target. You only require an infinitesimally tiny sample of the moonlight to gather the information that you need about either the Moon itself, or about the objects in your environment that interact with the moonlight.

What about a rainbow? It is easy to imagine that when you look at a rainbow you are seeing light itself in all of its various wavelengths. But in reality, you are seeing the combined effect of billions of raindrops which reflect, refract, and scatter sunlight, in all directions. When you look at a rainbow you are not seeing the light itself, but you are not seeing a real object, either. No matter how you think about it, the image of a rainbow is an illusion, and it is difficult to succinctly describe what it is that you are really looking at.

What about a laser beam? Again, it is impossible to see a laser beam directly as it moves through the air. You can only see a laser beam if you also have sufficient dust, smoke, or water droplet particles suspended in the air. These particles must reflect and scatter the light so that some small portion of it can reflect back and find its way to a pupil.

Imagine that you have an incredibly advanced, ultra-high speed video camera that allows you to capture and play back a video in extreme slow motion. If you had such a device, then you could also aim a laser pointer at some point in the distance and view the motion of the laser beam as it moves through the smoke or fog. From this, one might think that it is possible to directly observe the speed of the laser beam. But in this case, the apparent speed of the laser beam would actually be roughly half of the actual speed of the laser beam. This is because as the beam of light lengthens, the light must travel a progressively longer distance, both from the tip of your laser pointer to each reflecting particle, and from each particle and back to your pupil. It is impossible to directly see how fast the light is actually moving through the air. Instead, we need to rely on theory and mathematics in order to observe the speed of the light.

If the direction of the beam of light was reversed such that it was pointed towards instead of away from you, then the same beam would appear to travel instantaneously. No matter what the speed of the light really is, the speed of the visible beam that you see would appear to be infinite from your vantage point.

What about photons? Under low light and underexposed conditions, grainy, and noisy images are the rule. The tiny, dot-like objects that comprise an image are considered to be particles of light. These are extremely small, randomly scattered, and localized chemical reactions. What we see as a *photon* is just a leftover impression which remains long after the light has disappeared. The objects that we identify as *photons* are much like the footprints of an elusive creature that nobody has ever seen directly. The objects that we do see are not the light itself, nor are they the photons themselves. Considering the fact that photons are presumably point-like particles which always move at the speed of light, then it is not surprising that nobody has ever seen a photon directly.

The bottom line is that in order to see light, you also need to apply a model and a theory about what light is and a set of assumptions about how light behaves. Nothing about the nature of light is immediately obvious. Fortunately, as a species, we have never needed to understand light in order to use it effectively. We know that light helps us to see the objects that we are interested in, but when light itself becomes the object of interest, we find that its behaviour is strikingly unique. The behaviour of light is notoriously difficult to wrap your head around, both from the standpoint of special relativity and of quantum mechanics. To begin with, let's take a look at light through the lens of special relativity.

Leaving the Earth-Centred Inertial Frame of Reference

Imagine that you are stranded on a raft in the middle of the ocean. The sky is clear, the Sun is directly above you, and there is no land as far as the eye can see. You feel disoriented. You don't know where you are, or how fast you are floating, or in which direction you are floating. As you scan the horizon, you don't even know which direction you're looking in. This is the feeling of vertigo that we must experience whenever we leave our familiar Earth-bound frame of reference behind.

Physicists were startled when it was first confirmed in 1887 that the speed of light remains constant from the point of view of all observers. This finding might sound rather boring and innocuous until you begin to consider its implications. One important implication is that this makes it impossible to use the speed of light to determine how fast or in which direction you are moving. In other words, a constant speed of light represents a fundamental limit on our access to knowledge about motion. But this observation might also have been predictable, without any need for experiment, if only we could eliminate a few comforting, yet primitive assumptions.

Firstly, we would have needed to leave our Earth-centred frame of reference and forget about the idea that there is anything special about Earth. This abandonment is akin to opening up a can of worms. The idea that Earth is not the center of the universe is no longer a revolutionary notion, and yet it has proved difficult to fully embrace the logical consequences of this mundane fact.

It seems relatively clear now that Earth is not really stationary. For example, Earth is rotating, and as it rotates it is orbiting the Sun, the Sun is orbiting the center of the Milky Way, the Milky Way is moving towards the Andromeda galaxy, the Andromeda galaxy is moving relative to other galaxies, and so on, and so forth. But, for the most part, people still live their lives as though Earth is motionless and sitting at the centre of the universe. For most people, and for most of the time, our perception of reality is literally grounded on planet Earth, and it takes a lot of effort to temporarily break out of this mode of thinking. Even though we might know better on some level, we still regard the fundamental directions as east, west, north, south, up, and down, and we still assume that we are not moving, whenever we remain at the same location on Earth's surface.

Secondly, we would need to reject the idea that there must be some suitable substitute for our Earth-centred frame of reference. There is no objective way to select one particular inertial frame of reference over another. Nature has expressed no preference here. Not only is Earth not sitting motionless at the centre of the universe, but neither is the Sun and neither is the Milky Way. Nor is any other planet, star, galaxy, or intergalactic structure of any kind. Because all inertial frames of reference are inherently equal, any choice among them would be inherently arbitrary. So, while we might be most comfortable with our Earth-centered frame of reference, we must also abandon it without clinging to a similar surrogate if we are to understand how the universe really works.

The reason that I attempted to describe a feeling of vertigo in an evocative manner is to give you a way to see if you are on the right track. If you never experience this profound feeling of disorientation and bewilderment, then this necessarily means that these ideas have yet to sink in. Unfortunately, it is both easy, and common to accept or agree with a proposition without adequately processing it. If you do understand what this experience feels like, then rest assured that you are on the right track, and you can proceed. If you haven't yet experienced this feeling at least once, then you are not alone, and you are in good company. However, it is important to acquire a visceral feeling for what this all means, before moving forward.

Frames, Objects, and Observers

Let's begin with a very simple working hypothesis. Our working hypothesis is that how we choose to represent reality has no direct or immediate effect on that reality itself. Reality exists prior to our arbitrary systems of categorization, naming, and framing. While our modes of representation are variable, the reality that is referenced by our representations remains fixed. We may eventually find that this working hypothesis is flawed in some way, but as it comports with common sense, we should accept it until we find a compelling reason to reject it.

In order to gain a firm and clear grasp of frames, objects, and observers, and how they relate to one another, it is best to begin with a simple model with only two spatial dimensions, and no time. Before we can construct a two-dimensional coordinate system, we are presented with several arbitrary choices that must be made. We must choose where to put location zero, we must choose the scale of our units, and we must choose the angle of orientation. In other words, relative to the real terrain, we can slide our grid lines in any direction, we can expand, or contract the spacing between our grid lines to any size, and we can also rotate our grid lines so that they are oriented at any angle. We can change these three properties of the coordinate map in any way we choose to and obviously this will have no impact on the underlying terrain. The only real effect of our choices will be to nominally impact the way that we describe or represent the terrain.

As we vary the location that corresponds to zero on our grid, the locations of every point on the terrain will also *appear* to change, and in the opposite direction, but they won't really change. As we vary the scale of our grid, then the apparent location of every object will also change, unless it is located at zero. The distances between each object will also appear to change in equal proportion. If we make the grid units larger, then the objects, and the distances between objects, will appear to be smaller. If we make the grid units smaller, then the objects, and the distances between objects, will appear to be larger. As we vary the orientation of the grid by rotating it, the entire terrain will also appear to rotate, but in the opposite direction.

In this example the coordinate system/grid is the frame, the objects are those aspects of reality that we are adding to our map, and assigning coordinates to, and the observer is us because we are looking at this reality through its representation on a two-dimensional map. The main take home point here, is that the particular location that we choose to represent as *zero* is entirely arbitrary. The location of point zero is a feature of how our coordinate system is positioned only and not of the corresponding reality.

Because nothing about the terrain actually changes, as we make arbitrary changes to our coordinate system, it should be easy to identify certain invariant properties which would remain fixed, no matter which sort of coordinate system we construct. So, for example, if the distance between rock A and rock B is three times the distance between rock B and rock C, and if the angle formed between rock A, B, and C, is thirty-seven degrees, then everything that I have just stated about the three rocks will always be demonstrably true, no matter how our grid is positioned, oriented, or scaled. Both the angle and the relative distances are invariant properties of the geometrical pattern that is formed by the three rocks. This basic insight might not seem important at this stage, but it will come in handy in short order.

The Galilean and Newtonian Revolution

So far, our understanding of reality is primitive because we haven't yet considered the revolutionary insights of Galileo Galilei and Sir Isaac Newton. It is natural to assume that objects tend to passively maintain a fixed position on the ground unless there is some sort of active disturbance. Most of the objects on a map such as rocks, trees, mountains, roads, buildings, lakes, etc., maintain a fixed position, and so it seems reasonable to assume that anything that moves must be under the influence of some sort of energy. This understanding was good enough for our ancient ancestors and it is still adequate for daily life, but it is not sufficiently broad to assist us in making sense out of the behavior of celestial objects.

ORDINARY SPACE

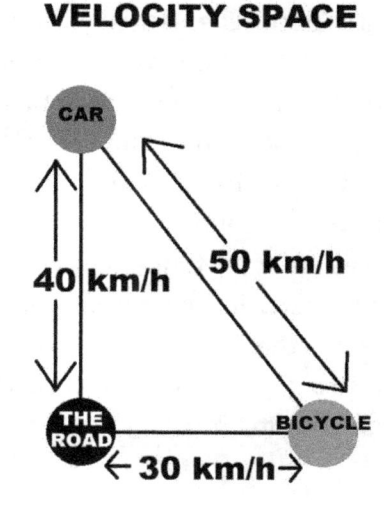

TULIP TREE
40 m
50 m
DAWN REDWOOD
GINKGO TREE
← 30 m →

VELOCITY SPACE

CAR
40 km/h
50 km/h
THE ROAD
BICYCLE
← 30 km/h →

In order to make celestial objects easier to understand, we will need a map of velocity space. Now, let's consider how we might construct a map of velocity space. Our grid will now be a three-dimensional lattice of cubic units, but we will also take the dimension of time into consideration, when thinking about how the frame of reference, the objects, and the observers are related. Taking time into consideration means that we must somehow represent the motion of objects in three-dimensional space over time on our grid. In other words, we no longer have a grid that gives us information about the position of objects. Instead, what we have is a grid that encodes information about the motion of objects. Because motion is the derivative of position over time, the new velocity map will also be the derivative of a three-dimensional position map that includes a dimension of time. In this new velocity map, the apparent location of an object represents its velocity. Any object that is not moving will be at location zero, and any object that is moving at a constant velocity will be at some position other than zero. The faster an object moves, the further its distance will be from zero.

The Galilean and Newtonian understanding of reality that was developed and adopted in the sixteenth and seventeenth centuries is surprisingly difficult to wrap one's head around, at first. What must first occur is essentially the replacement of the old position-based map with a new, velocity-based map, and the basic characteristics of the old map must then be transferred to the new map. Whereas, before we had assumed that objects have an inherent tendency to remain in the same fixed position, we now understand that objects have an inherent tendency to remain at the same fixed velocity. This inherent tendency is called *inertia* and Newton expressed this succinctly in his first law of motion. Newton's first law of motion states that every object will remain at rest or in uniform motion in a straight line unless compelled to change its state by the action of an external force.

Whereas, before we clearly understood that the position that we designate as *zero* is arbitrary, we now understand that the velocity that we designate as *zero* is also arbitrary. Prior to Galileo and Newton, we had assumed that our position-based grid did not itself move, but now we must discard the assumption of motionlessness. This realization opens up many possibilities for consideration, including the idea that Earth might orbit the Sun, as opposed to the other way around. We have just taken our old conceptual map and taken the derivative in order to add a new layer of abstraction. We can also see that our old conceptual framework only applied to a specific terrestrial case, and that our new conceptual framework applies much more broadly. Our new map will inevitably lead to new questions, new insights, and new paradoxes.

The location of an object or an observer on this three-dimensional velocity grid must have some special cosmic significance. Why else would there be a fundamental law of physics that dictates that everything must remain in a fixed location in velocity space, by default? Now that we recognize that velocities tend to remain fixed on this velocity map, we are faced with new questions about what is holding the objects in place, and about why they are being held in place in this new velocity space. Thought provoking questions such as these may be too deep to be resolved with any satisfactory answer at this point, but at least we now have a convenient way to visualise inertial motion. Not only do we now represent velocity as a position on the new map, but we also understand acceleration, and force, as the movement between positions on the new map. In other words, velocity is the new position and acceleration is the new velocity. Now that the velocity is represented with a location on our new three-dimensional map, there is no longer any way to represent position.

The key insight from this exercise is the fact that our choice of what counts as motionless or *velocity zero* on our map, is entirely arbitrary, just as location zero was arbitrary in our previous map. It is easy to underestimate how difficult it is to wrap one's head around this fact and to fully incorporate it into one's understanding. There is a persistent failure to account for both the fact that a zero velocity is arbitrary, and also for the fact that all velocity values are arbitrary. Extra care still needs to be taken so as not to confuse the arbitrary properties of our grid with the properties of reality itself. Just as it would not make sense to ask how far an object is, unless you also have a specific point of reference in mind, it also doesn't make sense to ask how fast an object is moving, unless you have a specific velocity of reference in mind.

Defining your Location in Velocity Space

We should have a clear way to describe the nature of what it is that is really fixed when you are locked into inertial motion, and you are not accelerating. What is it that I share with Earth when I am not moving relative to Earth? What is it that I share with all of the other objects and observers who are also motionless relative to Earth?

It is not actually a frame of reference that we share. Rather, it is a specific location on a three-dimensional velocity grid that we share. I say this, because the frame of reference is just an abstract means of orienting yourself and interpreting your observations. You can easily alter your frame of reference without changing your velocity, and you can easily alter your velocity without changing your frame of reference. So, for example, if I select Earth as an anchor for a frame of reference, and I also move relative to Earth, then I will perceive myself as moving. But I can also shift my preferred frame of reference, such that I am stationary, and it is Earth that is moving. The shifting of the frame of reference occurs purely in the mind.

However, there is a shorthand convention which holds that every observer is placed at location zero on a three-dimensional velocity grid, by default. Therefore, it would follow that each observer does have a single inertial frame of reference that is anchored to them. In other words, each observer is always at the center of the universe, and each observer is always motionless. I suspect that the logic behind this convention has something to do with explaining the strange behavior of light. It is okay to follow this convention, as long as it is always understood that the inertial frame of reference remains arbitrary and does not determine the objective reality. An observer may have a default inertial frame of reference, but this frame of reference can also easily be transcended. Choosing a location zero or a velocity zero is fundamentally no different from deciding when year zero should be. Because it is more convenient to go with the flow, I have been following the convention of assigning a default frame of reference to all observers, up until this point. However, going forward, I will not be using the phrase "inertial frame of reference." In its place I will be using the phrase *location in velocity space* because it is not really the frame of reference but the location in velocity space which remains fixed. The idea of a location in velocity space will be much more useful and much less confusing, but it will also be necessary to become more familiar in a practical way with the nature of velocity space.

I will also be advancing and advocating for the idea that just as your specific location in position space affects what light you are and are not able to see, so does your specific location in velocity space. When you change your location in position space, the light that you are able to interact with, and observe, shifts accordingly. Similarly, when you change your location in velocity space, the light that you are able to interact with, and observe, also shifts accordingly. There is no obvious explanation for why your location in velocity space has this effect of narrowly selecting only a tiny fraction of the light and rendering it visible. However, we also do not have any profound or deep explanation for why or how it is that your location in position space narrowly selects the light that you are able to observe. That is just how it is, and we may need to just accept this as a brute fact about reality. This interesting relationship to light may even be what defines or determines a location in velocity space. I only raise this point here to point out that there is yet another apparent parallel between position space and velocity space that is worth exploring.

To summarize, while it may appear that you are motionless because you are not moving relative to some other object (such as a vehicle or a planet), this is not really true. The deeper reality is that both you and the object that you are comparing your motion to are actually moving at the same speed and in the same direction. However, there is still no particular speed or particular direction that you are both moving in because the manner in which you describe your common location in velocity space also depends on where velocity zero is located. But regardless of where velocity zero is placed in velocity space, you will always find that you share a common speed and direction with that other object. Your relative motion with respect to that particular object is an invariant property which remains constant across all potential inertial frames of reference, as does the fact that you both share a velocity relative to everything else. What you share with that other object is a point in velocity space, and so the distance of that object from you in velocity space, is zero m/s. There is also something else which, in appearance only, must always maintain a fixed distance from you in velocity space, that is equal to 299,792,458 m/s. This is a fresh and revealing way of looking at the speed of light.

Relative vs. Subjective and Absolute vs. Objective

The main reason for relying on velocity space is to get around the strong and persistent tendency to confuse relative velocities with subjective velocities and objective velocities with absolute velocities. Velocities can be both objective and relative at the same time, and it is this fact which makes it possible to objectively map out the locations of objects and observers in velocity space.

The idea that all velocities are subjective just because they are relative is false. Unfortunately, this is no mere straw man argument.

Confusion between subjectivity and relativity is already baked into the language of special relativity because it is understood that an "inertial frame of reference" is both subjective and relative.

If the velocity of an object is subjective then this means that different people will disagree about where to place that object in relation to other objects in velocity space. If the velocity of an object is relative, then this means that different people will disagree about where to place the zero coordinates in relation to other objects in velocity space. So, subjectivity has to do with a dispute over where the object really is, while relativity has to do with a dispute over where the zero coordinates should go.

Establishing an Objective Orientation

True Distance vs. Frame-Dependent Displacement

I need to draw a sharp distinction between a robust property of nature and a relatively fragile concept. The meter is the standard unit of measurement for both of these quantities, but yet they are radically different. I will start out by defining what I call the "true distance" and then I will contrast this with what I call the "frame-dependent displacement."

True Distance

The robust property of nature that I have in mind is *true distance*. Questions about how large or how far away an object is are really questions about true distances. Examples of true distances would include the diameter or radius of an object, the distance between two objects at a particular instant in time, or the fixed distance between two objects which both share the same location in velocity space. This is our most essential and intuitive understanding of what distance is. True distance is closely related to but distinct from a relativistic concept known variously as the "proper length" or the "rest length." Unlike the proper length or the rest length, the true distance does not need to be measured in the rest frame only.

Concrete examples of true distances might be the distance around the equator, the distance between Earth and the Moon, or the distance between Chicago, and Toronto. What these distances all have in common is that they do not depend sensitively on where we decide to place our zero in velocity space. One would ordinarily say that these distances do not depend on the frame of reference of the observer because they were not derived from velocity. These distances are frame independent.

The true distance is what a tape measure is intended to measure. We also rely on rather sophisticated mental models which use visual angles as raw data and apply the rules of optics in order to draw inferences about true distances. We do not need to rely on either measures of time or of velocity in order to determine a true distance. The critical thing to note is that the true distance is absolute because it does not depend on any inertial frame of reference. True distances do not take either time or velocity into account in their calculation.

Frame-Dependent Displacement

We can use the stable visual angle of the Moon in order to determine that it maintains both a fixed distance from Earth and a fixed size. We are on firm ground, so far. But if we choose to make a trip to the Moon, then we will only be able to determine how much time our trip will take. As counterintuitive as this claim may sound, it is *not* possible to determine how long, or how far our trip to the Moon would be. Our trip to the Moon could cover any distance in any direction, depending on an arbitrarily chosen frame of reference. That is, the total length of our trip will be different, depending on where we arbitrarily decide to place the zero in velocity space.

This velocity-based distance will be called the frame-dependent displacement, which is simply the velocity relative to a frame, in a straight line path, multiplied by the amount of time taken. I could also have named this quantity the *velocitytime*, due to the manner in which it is calculated, but *frame-dependent displacement* is better for emphasizing the inherent fragility of this construct.

To make my point clearer and more intuitive, if you were to travel between two moving objects, then the length of the path that you take can be either much greater or much less than the true distance between the two objects. Whether the length of the path is more or less than the true distance will depend on the alignment between the direction of the path and the direction of the motion of the two objects in the chosen frame. If the traveller is considered to be at rest while moving between the two objects, then the total distance travelled will have to be regarded as *zero*. In other words, the *traveller* did not move but the other two objects did.

The Earth-Moon system is currently moving at more than 137 km/s relative to the center of the Milky Way. Meanwhile, if we place the Earth-Moon system at velocity zero, then the average speed of a typical space shuttle as it makes this trip, would be closer to 2 km/s. Therefore, if you use the Milky Way as an inertial frame of reference, then you would find that the total distance of a trip from Earth to the Moon would tend to be much longer than the true distance between the two objects. Most significantly, the more time that the trip takes, the longer the apparent path of the trip will be. More time equates to a farther distance because you are simply multiplying the velocity of the vehicle by the amount of time that the trip takes in order to get the distance. In other words, a slower rocket will result in a longer path taken, while a faster rocket will result in a shorter path taken.

To reiterate, unlike the frame-dependent displacement, the true distance between Earth and the Moon does not depend either on the velocity of the spacecraft or on the amount of time that the trip takes.

Frame-Dependent Displacement and Light

Because light moves at a speed which is so high that it may be regarded as infinite for most practical purposes, it follows that the frame-dependent displacement of light can normally be considered to be equal to the true distance. At an extreme speed of 299,792 km/s, any slight divergence between the true distance and the frame-dependent displacement would be negligible. However, the difference between the true distance and the frame-dependent displacement starts to become significant once we consider a frame of reference that moves relative to other objects at a speed that is approaching 299,792 km/s.

The distance between two mirrors which are facing each other, and which are locked into a fixed position can be objectively determined and this distance is frame independent. The distance between the two mirrors is a true distance because the mirrors both occupy the same location in velocity space. However, because some time does elapse as light reflects back and forth between the mirrors, it is natural to assume that we can use some information about distance, in combination with the information about the time elapsed, in order to calculate both the speed of the light, and the length of the path taken by the light. We are then faced with an important decision if we chose to make these calculations. Should we enter the true distance between the mirrors into our equation or should we enter the frame-dependent displacement into our equation? If we use the true distance, then our conclusion about the objective velocity of the light will be the same no matter where the observer is in velocity space. But if we use the frame-dependent displacement, then our conclusion about the objective velocity of the light will change, depending on where the observer is in velocity space.

It was a mistake to define the meter as the distance travelled by light in a fixed amount of time because this distance is merely a frame-dependent displacement. However, the frame dependent displacement of a photon is very different from that of a ball. The frame dependent-displacement of a ball depends on the distance between the ball and an arbitrary location zero in velocity space. The frame-dependent displacement for a photon also depends on the distance between the photon and an arbitrary location zero in velocity space. By convention, location zero is always considered to be the location of the observer in velocity space. Therefore, the frame-dependent displacement of a photon must always be exactly one light year for a time interval of one year.

While the frame-dependent displacement of every other object is variable, the frame-dependent displacement of a photon is fixed. But, in both cases, the frame-dependent displacement is a fictitious quantity. The frame-dependent displacement is not real. The frame-dependent displacement of a photon does not become real just because it is fixed. The fact that this quantity is fixed with respect to the photon makes it less real, not more real, than that of a ball. It is not wise to base a system of measurement on a quantity that is not objectively real.

According to SSL theory, the distance of the path taken by the light as it bounces between two mirrors cannot be objectively determined because this distance does depend on the frame of reference of the observer. The distance travelled by the light that we see is really a frame-dependent displacement and *not* a true distance.

In other words, what we can know for certain is the true distance between the two mirrors. We can also know for certain the amount of time taken by the light as it bounces between the mirrors. But what we *cannot* know is the distance of the path taken by the light as it bounces between the two mirrors, during that time. The distance of the path taken by the light in the nanosecond or so that the trip takes is actually everything from zero meters to infinite meters. If the frame-dependent displacement is represented as a vector, then this vector can point in all directions, and have all magnitudes simultaneously because our frame of reference is arbitrary and subject to change.

I maintain that Einstein was incorrect to assume that the distance of the path taken by a beam of light can be known. According to SSL theory, the distance of the path taken by a beam of light as it travels is purely a fictional construct with no basis in reality. Similarly, according to SSL theory, while we know the velocity of the manifest light, the latent light does not have any particular velocity. Furthermore, according to SSL theory, our knowledge is strictly limited to the true distance between the mirrors and the amount of time that the manifest light takes to travel between them.

Both Einstein's definition of simultaneity (from which he derives the concept of length contraction) and his light clock thought experiment (from which he derives the concept of time dilation) begin with the assumption that one can objectively determine the length of the path taken by a beam of light as it travels. Length contraction and time dilation are also both undergirded by the assumption that the velocity of light can be worked out from calculations that are based on the length of the path taken by the light (the frame-dependent displacement) in combination with the amount of time elapsed.

Special relativity is based on the equation below, which was presented by Albert Einstein in "On the Electrodynamics of Moving Bodies":

$$velocity = \frac{light\ path}{time\ interval}$$

Because both the length and the direction of the light path are aspects of its frame-dependent displacement, it will change depending on where we arbitrarily choose to place velocity zero in velocity space. In order for the observed speed of light to remain constant, Einstein assumed that the time interval must also change accordingly. For example, if the apparent light path shrinks, then the time interval must also shrink to the same degree. However, this is implausible because the arbitrary location that we select as velocity zero in velocity space should not have any impact on the physical reality. This is merely a property of our map and not the terrain. Therefore, it can only be the objective location of the observer in velocity space, and not their frame of reference, that has any real impact on their observations.

Following from the assumption that the frame-dependent displacement of light can be known, the light clock thought experiment asserts that, as the length of the path taken by the light increases relative to the true distance between the mirrors, then the amount of time that it appears to take for the light to traverse the mirrors, must also increase. Because speed is regarded as nothing more than distance divided by time, then in order for speed to remain constant, whenever the distance changes, time must change as well. This is the theoretical basis for time dilation, in a nutshell. One consequence of time dilation is that light is thought to follow a special rule that ordinary matter does not follow. According to this special rule, the speed of light is fixed, relative to the observer. Therefore, unlike with ordinary matter, the frame-dependent displacement of a pulse of light is regarded as a real, objective quantity.

SSL theory interprets the behaviour of light very differently. According to SSL theory, while the distance between the mirrors is a real, objective quantity, the distance of the path taken by the light is merely a pseudo-quantity with no objective basis in reality. The frame-dependent displacement of light is just as subjective and fictitious as the frame-dependent displacements of ordinary material objects are. I do have an alternative explanation for why the speed of light appears to be fixed relative to the observer, and for why its frame-dependent displacement appears to be absolute. What appears as a special rule for light that allows it to interact with the frame of reference of the observer, in an observer-centred way, is actually just an illusion based on the fact that light achieves a *supervelocity*, which means that light moves at a range of speeds and directions at the same time.

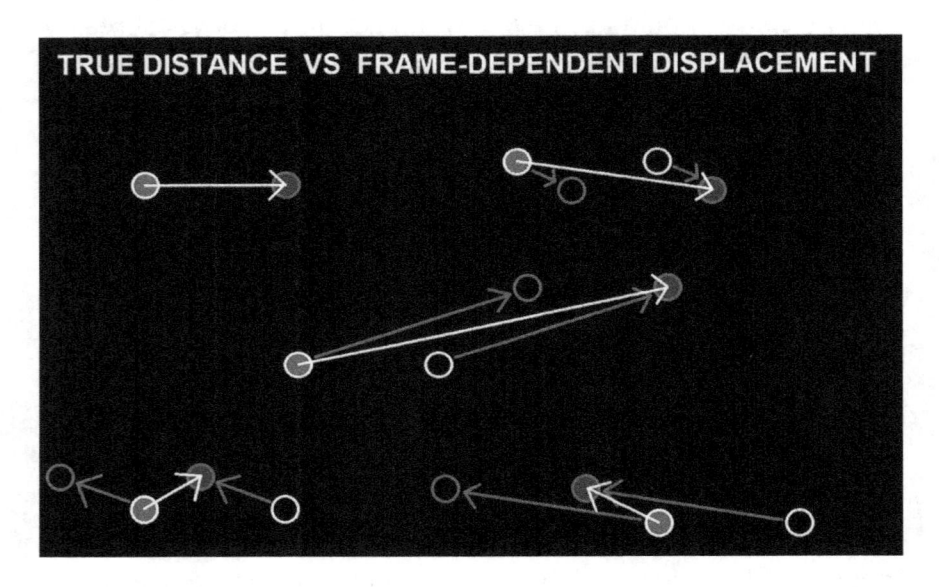

TRUE DISTANCE VS FRAME-DEPENDENT DISPLACEMENT

In the chart above, the white circles represent the location of two objects at *time one* and the grey circles represent the location of two objects at *time two*. The dark arrows represent the motion due to the frame over the time elapsed, and the white arrows represent the total frame-dependent displacement over the time elapsed. The distance between each pair of grey and white circles remains constant over time and this distance represents the true distance. The point of the image above is to show that the apparent distance covered by an object as it moves between the two points A and B depends on the relative direction and magnitude of the motion of the frame, as well as the amount of time taken, but yet the true distance remains the same, regardless of what the frame is doing.

There are at least two ways to ensure that the frame-dependent displacement is equal to the true distance. In the case where the circles A and B are stationary with respect to the frame over the time elapsed, and the dark arrows have a length of zero, then the frame-dependent displacement will be equal to the true distance. If the time taken by an object as it moves from A to B were zero, and its speed was therefore infinite, then the true distance would also be equal to the frame-dependent displacement. Another way to look at this, is to say that if the object moves at infinite speed between A and B, then it is at both locations at the same time, and so what you end up with will be a true distance.

If light is used to measure distance, then the information that we receive from the behavior of the manifest light does not indicate anything about what the latent light is doing. Instead, the information that we receive only tells us about where we are and how we are moving in relation to the other objects in our environment. Studying the time interval of manifest light cannot give us any information about true distances or about fundamental units of time. In other words, we cannot discover anything about how fast the light has actually travelled or about how far the light has actually travelled. The manifest light gives us no details about the path taken by the light.

How Does Light Achieve Supervelocity?

How Do We Integrate Multiple Observations?

According to conventional wisdom, the speed of light can be, and has been observed. This implies that specific velocities of light have also been observed. But according to SSL theory it is impossible to directly observe the full supervelocity of light. Similarly, according to quantum mechanics, it is impossible to directly observe a wave function. In the single photon double-slit experiment, a single photon passes through two slits at the same time, but in the form of a wave.

A photon is also able to interfere with itself while it is a wave. A photon eventually manifests as a particle, but only after it has been observed. Along the same lines, in the infamous Schrödinger's cat thought experiment, a cat is in a quantum mechanical superposition of both dead and alive while it is hidden inside a box. Once the box has been opened, and the cat has been observed, the superposition of dead and alive collapses, and the cat will appear to be either dead, or alive, but not both.

Just as with quantum mechanical experiments, what we measure when we measure the velocity of light is not really what we are interested in, and what we are interested in cannot be directly measured. Nobody has ever measured the velocity of light in its complete form. With a single photon, we only observe what appears to be a particle of light, but not its corresponding wave function. Likewise, at the macro scale, and from a particular location in velocity space, we can only observe what appears to be a wave of light. But we cannot observe or measure the entire superwave of light from any single location in velocity space, and therefore we cannot observe or measure the supervelocity of light directly. Instead, the complete supervelocity can only be inferred from multiple measurements taken from multiple locations in velocity space.

As I have indicated elsewhere, the length of the path taken by light is an inherently subjective measure of distance. With ordinary matter, the frame-dependent displacement is at all possible distances and in all possible directions at once, because the choice of inertial frame of reference is arbitrary. Similarly with light, all of the potential frame-dependent displacements are equally fictional and could each be visible from different locations in velocity space. Because velocity is the displacement over time, then over any common increment of time, this vast multitude of frame-dependent displacements becomes a supervelocity. But only a small subset of these velocities can actually be seen by a particular observer.

Because an observer can imagine and accurately model alternative coordinate systems, they are not really bound to the default perspective, and they have the capacity to integrate information from multiple perspectives. This capacity for integration across perspectives is precisely why there is no dispute as to which direction the light from a light bulb is travelling in. Rather, it is common sense that the light also fills up the room, and that other observers can also see the same glowing light bulb from other angles.

The distinction between a subjective frame of reference and an objective location in velocity space is important for the following reason. If a spaceship were to speed northwards, relative to me, at 200,000 km/s, then I should be correct in claiming both that the spaceship *appears* to be moving at 200,000 km/s relative to me, and that it *really is* moving at 200,000 km/s, relative to me. I should also be correct if I say that I am moving at 200,000 km/s southwards, relative to that spaceship. The idea that I am moving relative to the spaceship is just as correct and verifiable as is the idea that the spaceship is moving relative to me, in the opposite direction. Therefore, I am free to leave my frame of reference and occupy the frame of reference of the space ship without ever leaving my location in velocity space.

I can also reasonably take the perspective that the spaceship is stationary and that I am actually moving at 200,000 km/s, in the southward direction, while also seeing a second spaceship that is moving at 200,000 km/s, in the northward direction, relative to the "stationary" spaceship. Therefore, because I am able to integrate and make sense out of all of this information, I can also arrive at the correct understanding that the second spaceship should be moving at 400,000 km/s (faster than light speed), in the northward direction, relative to me. Simply by adding together two subluminal relative velocities, just as one would add together two relative distances, it is possible to create a superluminal relative velocity.

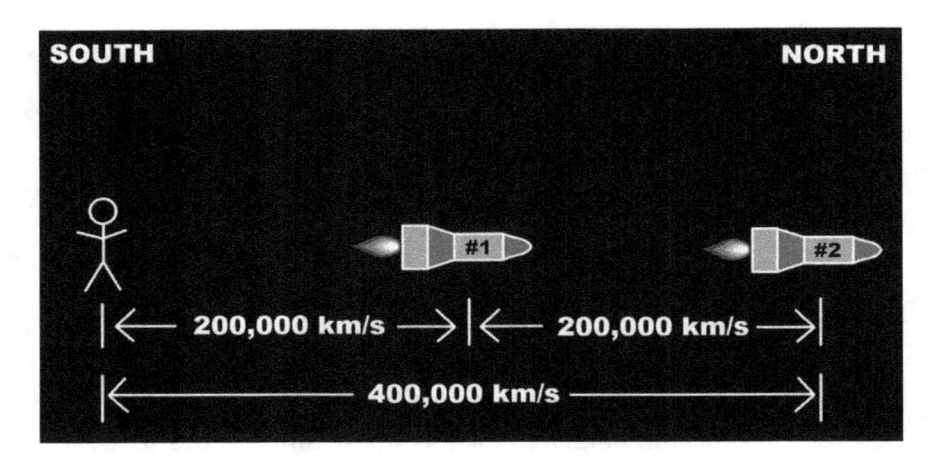

[Figure: Diagram showing an observer (south) and two spaceships #1 and #2 moving north, each at 200,000 km/s relative to the previous, totaling 400,000 km/s between the observer and spaceship #2.]

Furthermore, if I were to shoot a beam of light in the northward direction, then I should find it incredible that this same beam of light could be moving at 299,792 km/s, relative to me, while also moving at 299,792 km/s, relative to the first spaceship, and while *also* moving at 299,792 km/s, relative to the second spaceship. By integrating all three perspectives, what I end up with is the conclusion that the same beam of light is moving at three different speeds, and so it also occupies three different locations in velocity space, at the same time. These three different manifestations of the same beam of light would each be separated by a distance of 200,000 km/s, in velocity space. In other words, I end up with what is clearly a supervelocity because the light occupies multiple locations in velocity space at the same time. This is how it is that light can occupy a supervelocity and this perspective is perfectly consistent with the empirical observations that have been established to date.

I anticipate that most physicists will likely be shaking their heads in response to the mathematical assumptions about velocity that I have just made. This objection is based on the claim that spacetime is not *Newtonian*. Alternatively, this objection can be expressed as the idea that the universe is not *Euclidean*, or that reality does not conform to *Galilean invariance*, or that velocity space must apply *hyperbolic* geometry. What this objection boils down to is the idea that although

But Where is the Rainbow? 51

200 km/s plus 200 km/s does equal 400 km/s, 200,000 km/s plus 200,000 km/s does not equal 400,000 km/s. Furthermore, the sum of these two values must be less than 299,792 km/s, according to special relativity.

The question of how to add 200,000 km/s to 200,000 km/s within the paradigm of special relativity cannot be left unanswered, and so there must be some conventional formula that is used to integrate multiple perspectives. Such a formula would only need to satisfy two criteria in order to be consistent with special relativity. The first criteria is that the sum of the two speeds (with the same direction) must be greater than the larger of the two speeds on its own. The second criteria is that the sum must be less than 299,792 km/s. There is a large range between 200,000 km/s and 299,792 km/s and so there is plenty of room for a satisfactory answer. There is an infinite set of formulae which could satisfy these modest criteria, but there is no principled way to select any particular formula from amongst them.

A single, elegant formula has been adopted as the convention for adding together speeds in special relativity. According to the standard formula, 200,000 km/s plus 200,000 km/s is equal to 276,805 km/s. If we continue to add speed in increments of 200,000 km/s then the sum will eventually approach a limit of 299,792 km/s. This formula accomplishes exactly what it was designed to do, but so would an infinite number of alternative formulae. Towards the end of chapter five, I apply a cannonball thought experiment and an analogy with visual angles to bring some clarity to issues which I am only touching on here. Confusion between appearance and reality, translation between infinite and finite quantities, and a shallow and unreliable method for measuring velocity, all combine to make it difficult to interpret what is really meant by "speed" in special relativity. Hence, it is not the reality itself that is hyperbolic, but it is the accepted method of mapping between appearance and reality that is hyperbolic. In addition to being hyperbolic, this method has both incoherent and tautological aspects to it which will become clear before the end of chapter five.

If my frame of reference needed to be bound to my position in velocity space, then this would limit my ability to imagine, and to communicate how differences in velocity space could be added or subtracted from one another. This constraint would also make it impossible to map out and integrate different perspectives from multiple locations in velocity space. There would be no way to contemplate either superluminal velocities or supervelocities without the assistance of a velocity space framework. But because it is possible to separate an objective location in velocity space from an arbitrary and subjective inertial frame of reference, then it is possible to see both how an object could move faster than the speed of light, and how light could move at multiple speeds, simultaneously. Plotting out the location of light in velocity space for different observers also makes both superluminal speed and supervelocity easy to visualize.

What is Being Held Constant?

With the help of velocity space, the contradictory nature of the speed of light is much easier to see and much more difficult to ignore. One of the most common counterarguments that I have heard from physicists, is to deny that there is any contradiction or paradox displayed by the observed behavior of light. The commonly accepted belief is that the speed of light is a constant and so there is no contradiction or paradox that needs to be resolved. According to this perspective, all observers will agree on what the speed of light is. For those physicists who are inclined to follow this line of reasoning, I present a pair of analogies with money, and height, in order to make these contradictions as clear as they can possibly be made.

Analogy One: Money

Let's say that Alice has five dollars in her wallet, Bob has ten dollars in his wallet, and Charlie has twenty dollars in his wallet. Debra has exactly thirty-seven more dollars in her wallet than Alice does, exactly thirty-seven more dollars in her wallet than Bob does, and exactly thirty-seven more dollars in her wallet than Charlie does. So, how many dollars does Debra have in her wallet?

Analogy Two: Height

Let's say that Alice is five feet and six inches tall, Bob is five feet and nine inches tall, and Charlie is five feet and two inches tall. Debra is exactly four inches taller than Alice, exactly four inches taller than Bob, and exactly four inches taller than Charlie. So, how tall is Debra?

There can be no solution to either of these math problems, and if this is not immediately obvious, then it will become obvious upon attempting to find a solution. There is no single amount of money that Debra could have in her wallet, nor is there a single height that she could have, in order to make all of the contradictory statements about Debra, true. Debra would need to have three different amounts of money simultaneously or three different heights simultaneously. Debra's height and number of dollars would need to change, depending on who she is being compared to.

The idea that either Debra's height or the number of dollars in Debra's wallet can be regarded as a *constant*, is clearly absurd. Nobody would be tempted to say that the number of dollars in Debra's wallet remains constant at thirty-seven dollars more. Thirty-seven dollars more is not a specific value, it is merely a mathematical operation. Similarly, nobody would be tempted to say that Debra's height is a constant at four inches more. Four inches more than what? Four inches more than whom? Add four inches to what?

We cannot say that all observers agree on either the number of dollars that Debra has in her wallet, or on her height, because clearly all observers do not agree on any specific number of dollars or on any specific height. What all observers do agree on is a partial proposition which only sounds like a coherent idea until you attempt to objectively nail down its meaning. There is nothing objective, fixed, or finite about this pair of quantities that belong to Debra.

In just the same way that it is clear that Debra has a multitude of heights, but no height in particular, it should also be clear that light has a multitude of speeds, but no speed in particular. This multiplicity of velocity becomes just as obvious as the multiplicity of heights and dollars is when you try to objectively map out the velocity of light in velocity space.

But Where is the Restroom?

To drive home the absurdity of special relativity's account of the velocity of light, I have created the following dialogue which takes place in a restaurant. I call this dialogue, "But Where is the Restroom?"

> Customer (looking frazzled): Excuse me! Could you please direct me to the nearest restroom? It's an emergency!
>
> Server: Yes! The nearest restroom is exactly one hundred feet away.
>
> Customer (relieved): Great! Can you show me how to get there?
>
> Server: I've *just* told you exactly where the restroom is.
>
> Customer: You told me how far away it is, but I also need to know *how* to get to the restroom? So, can you give me more specific directions?

Server: Yes, one hundred feet is thirty-seven paces away. All you need to do is put one foot in front of the other. Thirty-seven times.

Customer (visibly uncomfortable): I understand that the restroom is thirty-seven paces away ... but thirty-seven paces away from ... where?

Server: Yes!

Customer: What do you mean, 'yes'?

Server: It really doesn't matter where *you* are. The restroom is thirty-seven paces away from anywhere.

Customer: So, if I were to walk down this hallway over here, for twenty paces, would I be closer or further away from the restroom?

Server: You would still be thirty-seven paces away.

Customer: Okay ... and what if I walk for twenty paces down that hallway over there?

Server: Still, thirty-seven paces away. Always thirty-seven paces away!

Customer: How is that ... possible? Does this restroom move around?!

Server: No? The restroom cannot move because it is *always exactly* one hundred feet away—that means its location can *never* change. No matter what.

Customer: I *really* need better directions on how to get to the restroom. Could you at least *point* to where the restroom is on this map of the restaurant, so that I can find it myself?

Server: What is a ... map? Such a strange concept! Nobody that I know has ever used a map before. The illustration on this 'map' is supposed to be a 'building' but it's just a bunch of lines drawn on a two-dimensional piece of paper. There is no *restroom* on this piece of paper. Therefore, I cannot point to it on this piece of paper. I'm sorry, but your 'map' doesn't make any sense to me.

Customer (looking around): Does this building even *have* a restroom? Because I am starting to suspect that it doesn't ...

Server: Oh yes! This building most *definitely* has a restroom. It is required to have a restroom, in fact. Otherwise, we would be in violation of the building code, and we wouldn't receive our permit.

Customer: Okay, and when the restaurant inspectors come to inspect your restroom, how do you assist them in finding it?

Server: We just tell them *exactly* where the restroom is. Eventually, the inspectors get confused and leave. (whispering) The inspectors wouldn't want to admit that they couldn't even find the restroom. *Nobody* has ever been inside the restroom, even though we *do* know exactly where it is.

Customer: What is the point of having a restroom, if it is inaccessible for all intents and purposes?

Server: We need to comply with the building code. (proudly) The distance to the restroom has been measured to twenty decimal places worth of precision. We are actually *more* certain about the precise location of the restroom, than we are about anything else!

Customer (suspicious): And how did you obtain this extremely precise measurement?

Server: We have an equation that *gives* us the answer! We plug in the size and the visual angle of the restroom into this equation here, and from that we can determine its proximity. You see, the ratio between the visual angle and the size is constant and so we also know that the proximity must also be constant!

Customer (now crossing legs anxiously): But how could the proximity of the restroom be constant? That doesn't make sense! Also ... I cannot wait any longer! I *need* to use the restroom, RIGHT NOW!

Server (condescending): Oh, but it *does* make sense to say that the proximity is constant, *if* you understand that the restroom doesn't *actually* 'move away' from you, as you walk towards it. You see, what actually happens, is that the restroom *shrinks,* relative to everything else, because everything else *grows*. Similarly, if you walk away from the restroom, then it will grow, as everything else shrinks! I know that this idea sounds counterintuitive, but the corresponding mathematics are flawless.

Customer (incredulous): But ... but ... how can I walk towards or away from the restroom if it is *always* exactly one hundred feet away from me and if *you* can't even show me where it *is,* on a map??! I need to see the manager!

(The manager arrives. He is rational and reassuring.) Manager: Sir, you just need to understand that what you call *walking* is simply a mathematical operation. I can perfectly describe what happens when you walk with just a few, simple equations. It will all make sense to you once you really understand the math.

Customer (exasperated): I understand that you are using mathematics, but you are *also* giving me directions in English. According to you, the restroom is thirty-seven paces away from me. But *also*, according to you, if I were to actually walk thirty-seven paces towards the restroom, then I *couldn't arrive* at the restroom! Doesn't that mean that the restroom *isn't* thirty-seven paces away? Could it possibly be, that what you call 'thirty-seven paces ...' isn't *really* thirty-seven paces???

Manager: From my perspective, I can walk thirty-seven paces, but you cannot. From *your* perspective, you can walk thirty-seven paces, but *I* cannot. So, what 'thirty-seven paces' means, depends entirely on perspective. In order to calculate a perspective, you need to use the right equations, and solve for gamma. While it is obviously impossible to say what 'thirty-seven paces' really means, I am absolutely certain that the restroom really is thirty-seven paces away. If you would like, I can show you the entry, in our dictionary, which currently defines 'thirty-seven paces away' as "the distance to the restroom from anywhere and the distance to anywhere from the restroom."

Customer: Never mind! It's too late now. But thanks for your help.

(Just to be abundantly clear, in the dialogue, the customer represents a philosophically minded student, the staff at the restaurant represent professors of physics, the map of the building represents a map of velocity space, and the restroom represents the velocity of light.)

Dynamically Defined Velocity: Light vs. Matter

A large, material object such as a ball in flight has a single, defined location in velocity space. However, the length of the path that is taken by a ball as it moves from A to B, is not defined. This means that the velocity of the ball is *predefined* but not *post defined*. It is predefined based on the location in velocity space of the objects that it has interacted with in the past. However, the fact that the same ball's motion can be observed from multiple frames means that its velocity is not post defined. The idea that the ball's motion is not post defined is another way of saying that its motion is relative to the frame of reference of an observer, as opposed to being absolute. The fact that the distance of the path of the ball and its velocity are not post defined means that it has all values simultaneously.

A similar but opposite phenomenon occurs with light because its velocity is not predefined, and the length of its path remains undefined. Hence, light moves at all relative velocities simultaneously. But at a later time, a small subset of velocities is selected by the observer based on the observer's location in velocity space. Hence, the velocity of the light is not predefined, but it is post defined. If we focus on the predefinition, then light appears to be less constrained than ordinary matter, but if we focus on the post definition, then light appears to be more constrained than ordinary matter.

We can model how predefinition and post definition work with an example of the ripples that expand in a pond after a pebble is dropped into it. The location of the centre of the concentric circle-shaped ripples is *predefined* by the location where the pebble landed. The energy is then transferred from the pebble to the surface of the water. The center of the concentric circles also remains fixed, relative to the medium, before the ripples interact with anything else. When the ripples do interact with an object that is floating on the surface, only a small part of the wave front interacts with that object. The direction of the motion of the wave relative to the object that it transmits energy to, is therefore defined by the relative location of that object. Hence, the direction or angle of the wave front is post defined by the location of the object that it transfers its energy to.

Similarly, the direction and length of a direct line from a source of light to the pupil depends entirely on the relative location of the observer. Hence, the properties of this line of sight are not defined until the location of the observer has been determined. If you remove the observer, then you also remove the line of sight. We can describe this situation in the abstract by saying that the line of sight is not predefined, but that it is post defined. If it had been defined at the source then it would have been predefined, but because it is defined at the pupil then it is post defined.

I had originally constructed the velocity matrix below in order to juxtapose the behavior of a photon (light) with the behavior of a marble (matter). This was one of the insights that eventually led me to discover SSL theory. The four interacting variables that I considered while constructing the matrix were subjective vs. objective, potential vs. actual, direction vs. speed, and light vs. ordinary matter. At that time, I had not yet considered the possibility of a supervelocity. The ones in this matrix indicate that there is only one possible value that could be obtained, and the infinity symbols indicate that there are an infinite number of possible values that could be obtained. The locations of the ones in this matrix make it clear that while the speed of light is subjectively *more* constrained than the speed of ordinary matter, the speed of light is objectively *less* constrained than the speed of ordinary matter.

		MARBLE		PHOTON	
		Actual	Potential	Actual	Potential
OBJECTIVE LOCATION IN VELOCITY SPACE	SPEED	1	∞	∞	∞
	DIRECTION	1	∞	∞	∞
SUBJECTIVE RELATIVE VELOCITY	SPEED	1	∞	1	1
	DIRECTION	1	∞	∞	∞

Manifest Light vs. Latent Light

While it is commonly understood that the vast majority of ambient light is never observed, I am introducing a novel sense in which only a small fraction of the light that is present in the environment can actually be seen. In addition to the limits on observation which are already set by the spectral bandwidth of visible light, the position of the observer, and the division of light into discrete photons, SSL theory posits one further limitation which is tied to the location of the observer in velocity space.

Every observation is the end result of some selection process. There can be no observation without a corresponding selection process. To observe everything is to observe nothing. To observe something is to filter out everything else.

It is a widely acknowledged fact that human beings are only capable of perceiving a small sliver of the available electromagnetic spectrum. We refer to the light that we can see as *visible light* and this part of the spectrum corresponds to wavelengths which are between 370 and 710 nanometers or to frequencies which are between 810 and 420 terahertz. The vast bulk of the electromagnetic spectrum is invisible to the naked eye, but we do have other ways of detecting invisible light such as gamma rays, x-rays, ultraviolet, infrared, microwave, and radio waves. The visible light is manifest while the rest of the light is latent. The selection process in this case involves complex chemical structures, such as rhodopsin, which can only absorb electromagnetic energy from wavelengths that are in the visible range.

It is also a widely acknowledged fact that a brightly glowing object emits light in all directions, and that most of that light will never be observed. This is just common sense, owing to the fact that a pupil is obviously a very small target, and so the vast majority of the light that could have been seen, will never be seen, merely because it was not directed at any pupil. Because we understand this common sense fact about optics, there is no dispute about which direction the light from a light bulb is travelling. The light can travel in all directions simultaneously. An observer can see that the light travelled in a straight line from the source and to their pupil, but they can also acknowledge that the same light also travelled in other directions, even though they are reasoning about light that they cannot see. Had the observer been in a different location, they would have still been able to see the same light, but the light would have appeared to be coming from a different direction. The light that is directed at a pupil is manifest whereas the rest of the light is latent. The selection process in this case involves a pupil which only accepts light that happens to be directed at it from in front of the eye ball.

It is also a widely acknowledged fact in quantum mechanics that as a photon travels it does so as a probability distribution that takes the shape of an expanding wave. When the photon is observed it will appear in a very specific location, but it could have just as easily manifested in any other location in a larger probability distribution. Upon observation and/or measurement of the photon, a superposition of potential eigenstates is reduced to a single eigenstate (a clearly defined property). The entire wave function and the corresponding superposition is understood to be a real object even though only a single position can actually be observed for any given experiment. The precise location where the photon actually appears is manifest whereas the rest of the wave function is latent. The selection process, in this case, is mysterious.

According to SSL theory, latent light also takes the form of light that travels at speeds which are either faster or slower than 299,792 km/s, relative to the observer. Only the light that travels at or near 299,792 km/s relative to the observer is manifest and can therefore be detected. Because different observers can occupy different locations in velocity space, light that is manifest to one observer will be latent to another observer and vice versa. The fundamental error that creates confusion is the assumption that the light that is manifest to one observer is the "same light" that is manifest to another observer who is in a different location in velocity space. In reality, the same light is not seen by both observers and the manifest light that really is seen by both observers is actually moving at different speeds. The effect of this error is similar to the illusion that is created by a hologram when it is assumed that both eyes are viewing the same object but from different angles. The light that is manifest is only the light which has been selected from the available latent light by virtue of the specific location of the observer in velocity space. The rest of the light which is latent will not interact with or be seen by the observer, but some of that latent light may still interact with and be seen as manifest light by a different observer who is in a different location in velocity space. The selection process in this case is also mysterious, but at least it produces predictable results.

Integrating Three Critical Distinctions

I have previously identified three interconnected distinctions here.

1. Breaking the term *inertial frame of reference* into two related concepts. Here, I am drawing a necessary distinction between location zero in the coordinate system that is applied and the location of an observer in velocity space: 1. The configuration of the coordinate system is relatively subjective, variable, and abstract, while 2. the location in velocity space is relatively objective, fixed, and concrete.

2. Breaking the term *distance* into two related concepts. Here, I am drawing a necessary distinction between the frame-dependent displacement between a point of departure and a destination and the true distance between those two points: 1. The frame-dependent displacement is relatively subjective, variable, and abstract, while 2. the true distance is relatively objective, fixed, and concrete.

3. Breaking the term *light* into two related concepts. Here, I am drawing a necessary distinction between the manifest light that is only visible to observers from within a specific location in velocity space and the latent light which is not visible and constitutes the rest of the light: 1. The manifest light must have a relative speed of 299,792 km/s, which is specifically tailored to the location in velocity space of a particular observer, while

2. the latent light occupies the remaining locations in velocity space. The observer always occupies the average location of the manifest light in velocity space (the centre of the bubble).

Now that we have the concept of velocity space, the concept of true distance, and the concept of latent light, we can assemble an objective model of what is really happening with light in a way that isn't helplessly dependent on a subjective and arbitrary perspective. While it does appear as though I am adding theoretical complexity to the subject matter in drawing these three distinctions, the end result of adding this complexity will be to arrive at a far simpler and more coherent theory.

Visualizing Velocity Space, Dark Matter, and Dark Energy

Through understanding velocity space as something objective which can be mapped out and visualized, the true character of the velocity of light can be made apparent. What is needed is a way to fully capture and characterize the fundamental discrepancy that is both implied and apparent whenever the velocity of light is measured or defined. This exercise will reveal both the depth of the reconciliation that is required and the inadequacy of Einstein's proposed solution. It will become clear to anyone who would otherwise have doubts that the observed velocity of light does actually depend entirely on the velocity of the observer. The use of velocity space permits both a mathematical description and a vivid visual representation of concepts which are notoriously difficult to convey with words.

In addition to light, other aspects of the universe will also take on a new appearance when viewed through the lens of velocity space. A velocity map will also give us a new and improved way to understand dark matter and dark energy, in addition to stellar, and galactic dynamics, in general.

I will first illustrate the concept of velocity space with a simple example. Imagine three cars driving on a straight section of highway. One is in the right lane travelling at 100 km/h, another is in the middle lane travelling at 110 km/h, and a third is in the left lane, travelling at 120 km/h. If we were to place these three cars on a map of velocity space, then the cars would be represented as points that are arranged in a straight line because all three cars are travelling in the same direction but at different speeds. Only the direction that each car is traveling in, and its speed are relevant. The fact that the cars are in different lanes does not matter, nor does it matter which order the cars are in, or how far apart they are from one another. The distances between each car on the velocity map would be ten km/h.

The Solar System in Velocity Space

To see why velocity space is useful in physics and cosmology let's begin with the solar system. You have likely seen diagrams of position space for our solar system many times and while a velocity map may look similar, it is radically different. In the two-dimensional velocity map below, I have included the Sun at the centre. The Sun occupies velocity zero in the grid below, just so that we can have a nice, symmetrical image. The white circles represent the orbits of Neptune, Jupiter, Earth, and Mercury, respectively.

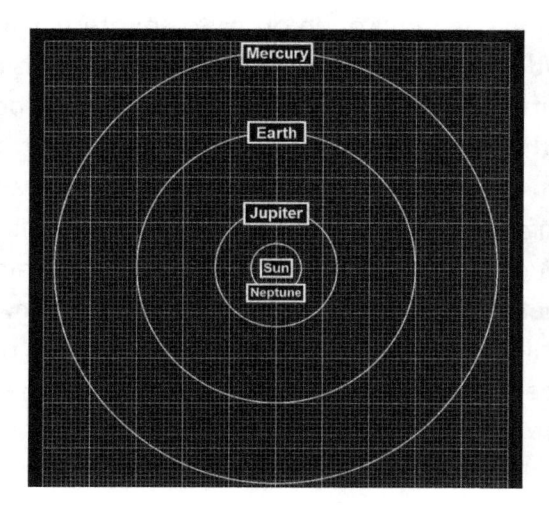

The most striking feature of this velocity map is the fact that the sizes of the velocity orbits are arranged in the reverse order. The planet with the largest position orbit has the smallest velocity orbit and vice versa. This inversion is due to the fact that planets which are further from the Sun orbit at a much slower speed than do planets that are closer to the Sun. Neptune orbits at a speed of 5.43 km/s, in all directions within the two-dimensional orbital plane. Jupiter orbits at a speed of 13.07 km/s, Earth orbits at 29.8 km/s, and Mercury orbits at 47.9 km/s, on average. It is also interesting to note that the circumference of Earth's orbit in velocity space is 187 km/s. The circumference of Earth's orbit in velocity space is a measure of the degree to which the Earth changes its velocity relative to the Sun, over the course of a year. The circumference of Earth's orbit in velocity space is also a measure of the total acceleration and/or force that Earth experiences each year, as a result of the sun's gravitational influence. What is recognized as Earth's orbital velocity (29.8 km/s), is also the radius of the Earth's orbit in velocity space. Both the radius and the circumference of a planet's orbit in velocity space are invariant properties which do not depend on an arbitrary frame of reference.

The simple explanation for why more distant planets move more slowly is that they feel the pull of the Sun's gravity more weakly, and so less speed is required in order to counterbalance the force of gravity, and to achieve a stable equilibrium. There is also another important difference between the features of the position orbit and that of the velocity orbit. At any given moment, the location of a planet in velocity space will be ninety degrees away from its corresponding location in position space. This happens because the direction of motion of each planet is always at a right angle from the directions that point towards and away from the Sun. The planets are forever sliding sideways at a right angle, and so their location in velocity space is always ninety degrees ahead of their location in position space. (I note that in reality the position orbits of the planets are not perfectly circular, and neither are their corresponding velocity orbits, but I have represented the velocity orbits as perfect circles in order to save time and to grease the wheels of communication.)

Stellar Velocity and Dark Matter in Velocity Space

The velocity map concept really becomes interesting once we start to look at how velocities are arranged at the galactic level. In a typical spiral galaxy, every star moves at roughly the same speed relative to the supermassive black hole that inhabits its centre, irrespective of the distance between the star, and the supermassive black hole. That speed is generally close to 240 km/s. If you want to get a sense of how fast 240 kilometers per second is, this is about 1,000 times the speed of a bullet.

The image below depicts roughly what a velocity map would look like for a typical spiral galaxy such as the Milky Way. As you can see in the velocity map below, every star in the entire galaxy occupies approximately the same velocity orbit. The image below is not as visually appealing as a standard image of a grand spiral galaxy, but its relative simplicity is astounding.

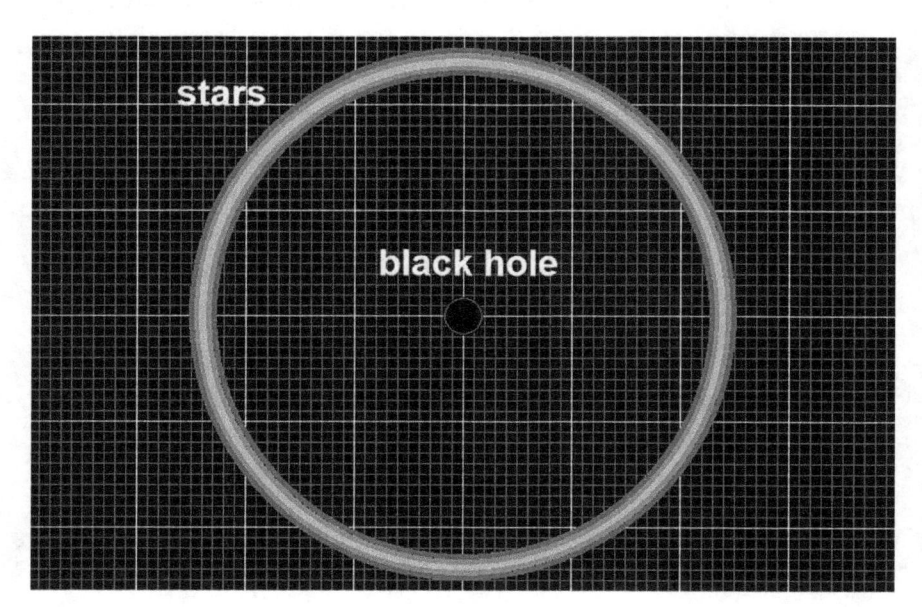

If you did not already know that stellar orbital velocity remains nearly constant at all distances from the centre of the galaxy, then this fact should be difficult to accept. What was originally believed and expected prior to the 1970s was that the stars in galaxies would follow a similar pattern to that of the planets in the solar system. Stumbling upon the inescapable conclusion that stellar dynamics are radically different was responsible for a genuine crisis in astrophysics.

There are two possible ways to respond to the shocking finding that the speed of stars remains stable throughout the galactic plane. One possibility would be to conclude that our current understanding of the mechanics of gravity is inadequate and that, by extension, Einstein's theory of general relativity is incomplete. In other words, we could conclude that general relativity does not apply at the galactic scale. The other possibility is to introduce a novel *fudge factor* in order to balance the equations out. A *fudge factor* is an ad-hoc quantity that is artificially introduced into a mathematical model in order to bring it into alignment with observations which would otherwise be incongruent, surprising, and inexplicable. You can think of a fudge factor as the scientific equivalent of duct tape.

In the end, the astrophysical community settled on the second option and a novel fudge factor was introduced into the equations. This fudge factor is now called *dark matter,* on the dubious assumption that there exists additional matter in the galaxy which we have no other way of detecting, except through its ability to modify the orbital velocity of stars. For a fudge factor, dark matter is absurdly substantial, and influential. Dark matter is thought to constitute roughly eighty-five percent of the total matter in the Milky Way galaxy. It is difficult to overstate how embarrassing and how humbling an error of this magnitude really is. Calling this fudge factor "dark matter" accomplishes little, other than to temporarily sweep a massively inconvenient fact under the rug, and to carry on as though everything is still in order. It still remains possible that dark matter really is a mysterious new form of matter, but we don't yet have any good reason to assume that it is matter at all.

All that we really do know, is that the stars found at the outer reaches of spiral galaxies move much faster than is physically possible, according to general relativity as currently formulated. Stars that orbit at an average distance from the galactic centre also move much faster than they should be able to, but the discrepancy between theory and observation is not nearly as pronounced for them. What is most fascinating about the overall pattern in the velocities of stars is how remarkably stable and consistent their speeds are. It is almost as though there is some ideal speed that the stars want to move at, and that this ideal speed is somehow a property of a particular galaxy.

Dark Energy and Light Bubbles in Velocity Space

Now, we will zoom out by a factor of 100,000, to see how the universe looks in velocity space. According to the current state of knowledge, once you reach a scale that is as large as the entire known universe (which is several billion light years across), then the velocity map, and the position map, begin to mutually resemble one another. The location of each particular galaxy in velocity space corresponds very closely with its location in position space. This is because at a large scale, the universe is expanding in an even, and consistent way. The further apart any two galaxies are, the faster those two galaxies will be moving apart from each other.

It was first observed by Edwin Hubble in 1929, that the galaxies are drifting apart from each other in a fairly regular and uniform manner across vast reaches of space. This was the observation that led to the establishment of the big bang theory.

In 1998, it was further hypothesised that the universe appears to be accelerating in its rate of expansion, based on the observed brightness, and red shift values of type Ia supernovae. With no explanation for the mechanism that generates this mysterious force, a new fudge factor was created, and *dark energy* was the label that was assigned to it. It is believed that most of the energy in the universe is dark energy, under the assumption that dark energy is actually a form of energy. Dark energy behaves like an extremely diluted sort of antigravity which permeates all of space and manifests as a repulsive force. Interestingly, now that dark energy is being used to explain the current rate of the expansion of the universe, the big bang theory has yet to be discarded as redundant.

The trouble that I have with the terms *dark matter* and *dark energy*, is that they do actually convey a false impression that each respective phenomenon has already been explained and understood well enough to assume that they are actually forms of matter and energy, respectively.

Because the universe is currently expanding in an evenly distributed manner, a three-dimensional position map and a three-dimensional velocity map should bear a close resemblance to each other at large scales. The same pattern will also hold true for two-dimensional cross sections of position space and velocity space. The image below shows eight hypothetical galaxies which are separated by vast distances in both position space and velocity space.

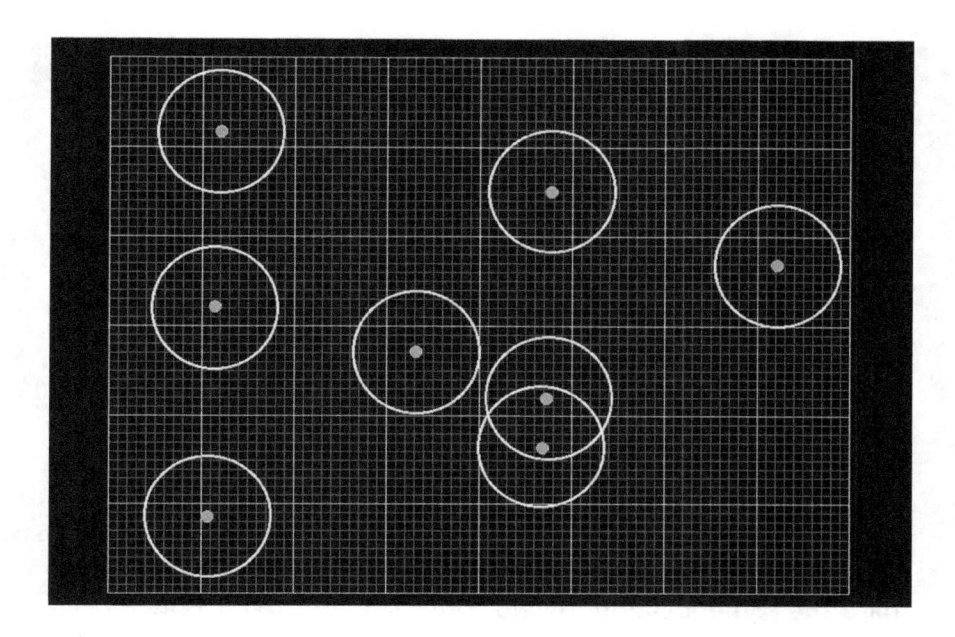

To get a sense of the scale, the closest pair of galaxies represented above are about 250,000 km/s apart in velocity space. The most distant pair of galaxies are about 2,500,000 km/s apart in velocity space.

To reiterate, velocity space is essentially the same type of space as ordinary position space, except that it is the derivative of position space over time. When translating from position space to velocity space, position becomes irrelevant, velocity becomes position, acceleration becomes velocity, jerk becomes acceleration, snap becomes jerk, and this pattern continues along the infinite series of derivatives. This shuffling occurs because velocity is a measure of the change in position over time, acceleration is a measure of the change in velocity over time, jerk is a measure of the change in acceleration over time, and snap is a measure of the change in jerk over time. Interestingly, just as Pythagorean theorem can be used to calculate distances in position space, it can also be used to calculate distances in velocity space, which is extremely convenient. The velocity space that I have envisioned is Euclidean. However, velocity space does not need to be Euclidean in principle. I prefer to work with Euclidean geometry because it is a reliable and stable system to begin with, as a default representation of reality. At the very least, Euclidean geometry is approximately accurate, even if it turns out that it is not perfectly accurate, or always accurate. Euclidean geometry is far simpler and more intuitive than the modified alternatives that have been developed in the twentieth century. Having said that, non-Euclidean geometries, such as hyperbolic or hypersphere geometries, can also be represented with velocity space.

Under a framework of velocity space, Newton's first law of motion would be translated into the claim that all objects must remain in the same location in velocity space, unless acted upon by an external force. All objects that are not moving relative to one another would share the same location in velocity space, no matter how distant they are. The speed at which any two objects are moving, relative to one another, is also the distance between them in velocity space. The specific direction in which two objects are moving apart from or towards one another, also corresponds perfectly to the direction of the distance vector that connects them in velocity space. However, it is not possible to tell whether two objects are moving towards, or away from, one another, merely by viewing them as points in velocity space. Distinguishing between increasing and decreasing proximity would require additional information about the relative position of the two objects at a particular moment in time.

Einstein's theory of special relativity can also be translated into the language of velocity space. When translated, Einstein's first postulate of special relativity would state that all fixed locations in velocity space are equivalent and cannot be distinguished based on the laws of physics. Einstein's second postulate of special relativity would state that manifest light always occupies a spherical bubble with a fixed radius of c in velocity space and with the observer always at the centre of this bubble. Hence, the observer cannot use the relative velocity of manifest light in order to determine their location in velocity space.

The white circles that you see around each of the eight galaxies in the above image are two-dimensional representations of light bubbles. Each light bubble shown above would be the default light bubble for the inhabitants of each of the eight galaxies. Wherever an observer travels in velocity space, that observer's light bubble must also move along with them. The only way that it would be possible for the light bubble of an observer to significantly deviate from their default galactic light bubble, would be for the observer to achieve an extremely high velocity relative to the rest of their galaxy. Much like a shadow, or a reflection, the location of the light bubble appears to track and mimic the location of the observer in velocity space.

By contrast, with ordinary objects such as planets and stars, their locations in velocity space do not vary with, or depend on, the location of any observer in velocity space. Dependence on the location of the observer in velocity space is the crucial difference between the apparent behaviour of light and the behaviour of everything else.

Any two observers who are located at different points in velocity space will each have light bubbles that surround them, but which do not overlap, except for a circle in velocity space that exists half way between them. This circle would represent the relatively narrow range of potential locations in velocity space that manifest light could hypothetically occupy, according to the perspectives of both observers. As the observers move further apart in velocity space, the circle of overlap will shrink until it becomes a point. At this point, both observers could see still see some manifest light moving at the same speed but in opposite directions relative to the observer, and the two observers would be moving apart from each other at twice the speed of light. Once the observers begin to move apart from each other at a rate faster than 2c, then there will no longer be any objective overlap between their observations of the behaviour of light.

The big question that divides special relativity from SSL theory is: how is it possible to reconcile these wildly divergent accounts of reality that emerge from observers in different locations in velocity space? The fundamental mistake which discourages inquiry into this question is to take the idea that the speed of light is *fixed*, merely because it is a constant according to special relativity, and to assume that the corresponding light bubble must also occupy a fixed position in velocity space. The description of the speed of light as a constant is a superficial and misleading linguistic convention which betrays the objective reality of light. The assumption that light has a single speed is clearly an error, and explicitly mapping the locations of the manifest light in velocity space makes this error even more conspicuous.

Relative Velocity and Simultaneity

In his seminal work "On the Electrodynamics of Moving Bodies," Einstein expresses the following thought process: 1. We must rely on the frame-dependent speed of light in order to determine that two different events, which are separated by a vast distance, are simultaneous. 2. We must establish that two different events that are separated by a vast distance are simultaneous, in order to determine the distance between them. It is important to establish simultaneity because if the two events were not actually simultaneous, then our inertial frame of reference will affect our estimate of the distance between them. From these two premises, Einstein concludes that because the speed of light is relative to the inertial frame of the observer, simultaneity in time must also be relative to the inertial frame of the observer. Einstein further concludes that because simultaneity is relative to the inertial frame of the observer, the distance between two simultaneous events must also depend on the inertial frame of the observer. The overall picture of Einstein's reasoning can be represented as the causal chain below:

Inertial Frame –> Relative Speed of Light –> Simultaneity –> Distance

I sympathize with Einstein's recognition of the difficulty of determining the simultaneity of events that take place, vast distances apart. It is strange to even consider what global simultaneity at an astronomical scale would mean or how it would ever become a useful concept. This construct would also be appreciably difficult to measure. However, contrary to Einstein's assumption, we do not need to rely on any absolute concept of global simultaneity in order to measure vast distances accurately because there are more direct ways of measuring distances which do not require the use of synchronized clocks and light signals.

For example, distances can also be measured by taking advantage of parallax, blurriness, and focus (resolution), visual angle, standardized sizes, and measuring devices (such as a ruler), or on the product of velocity, and time. I have just listed five different ways of measuring distance, but this was not intended to be an exhaustive list. So long as our conclusion about a distance does not depend on the method that we use to measure it, then we know that we have a robust, and useful concept.

If SSL theory is accurate, then using light signals to synchronize clocks, and then using those synchronized clocks to measure the time intervals of light pulses would be a relatively unreliable method for measuring distance. To put it simply, the combined use of clocks and light signals would end up conflating latent light with manifest light.

I will now reproduce Einstein's "Rigid Rod" thought experiment, in order to compare SSL theory with special relativity. Let's say that you have an observer that is halfway in between Clock A and Clock B and that this observer shares a location in velocity space with the two clocks. We would ordinarily say that the observer is stationary, relative to the two clocks, or that the observer, and the two clocks, are all in the same inertial frame of reference. In this case, if the observer receives a light signal from both clocks at the same time, then it is reasonable to conclude (based on the shared frame of reference and the equal distance of the two clocks) that the two clocks both emitted their signal simultaneously. However, according to Einstein, the situation is different when the observer is moving, relative to the two clocks.

Let's say that a second observer is moving from Clock A to Clock B at half of **c**, or 149,896 km/s. When the second observer reaches the midpoint between Clock A and Clock B, both clocks emit a signal simultaneously, from the perspective of the stationary observer. In that case, the moving observer will actually receive the signal from both clocks simultaneously, at the moment when they have travelled seventy-five percent of the way to Clock B. Both of the signals that the moving observer receives would have travelled at 299,792 km/s, relative to the moving observer. However, the signal that the moving observer receives from Clock A would have travelled at 449,688 km/s, relative to the two clocks, and this signal would have been invisible to the stationary observer. Meanwhile, the signal that the moving observer receives from Clock B would have travelled at 149,896 km/s, relative to the two clocks, and this signal also would have been invisible to the stationary observer. Both observers would have also received their signals from the clocks at the same time, but while at different locations.

WHAT THE "STATIONARY" OBSERVER SEES

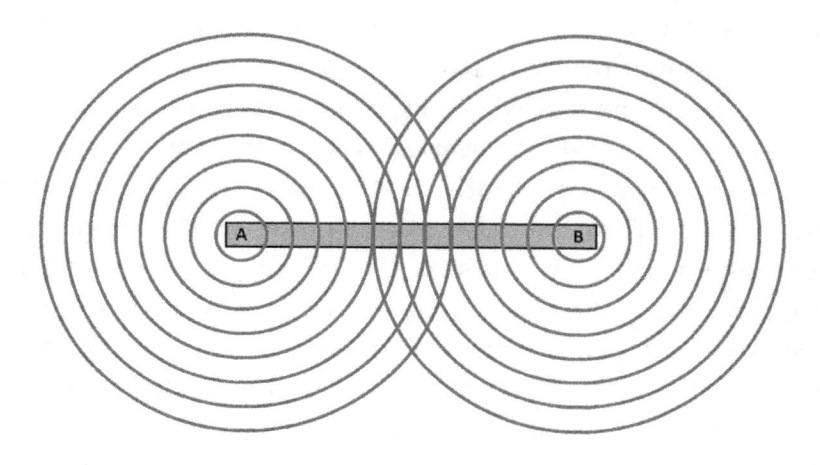

WHAT THE "MOVING" OBSERVER SEES

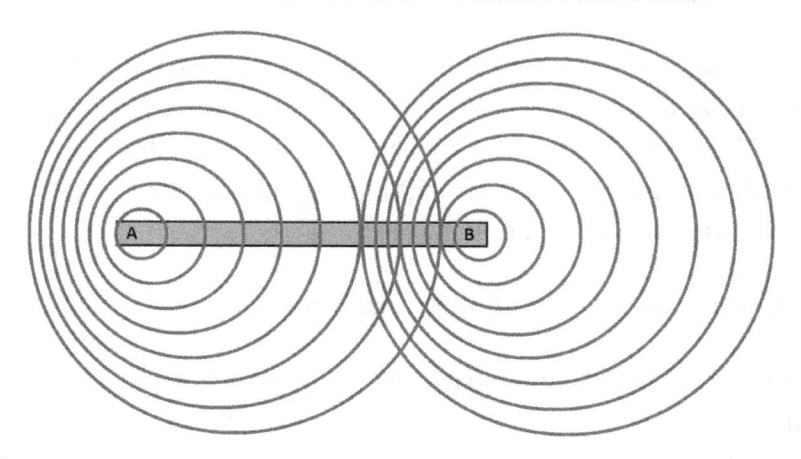

Because the moving observer receives the two signals at the moment when they are seventy-five percent of the way from Clock A to Clock B, one might assume that it is reasonable for the moving observer to conclude that the two signals had been emitted at different times. In other words, the moving observer will not perceive the two signals as simultaneous. But while it is true that the moving observer is much closer to Clock B at the moment when they receive both signals, this is not an accurate portrayal of how the situation will appear to the moving observer. The moving observer will actually interpret themselves as stationary because they are in inertial motion, and they will likely perceive the clocks as moving. If the observer interprets themselves as stationary and the clocks as moving, then they will also interpret the signals as having been sent simultaneously. From this perspective, it makes sense that the clocks have both moved in the time since the light signals were sent. From the perspective of the moving observer, both clocks will also appear to be exactly the same distance away because they won't be able to see where the clocks currently are, but only where they were when they emitted their signals. Every clue that the moving observer will have about the two clocks will tell them that they are both the same distance away, and this could include their apparent position, their visual angle, and the intensity of the signal.

The only noticeable difference that the moving observer will be able to perceive between the signals from each clock, is that the signal from Clock B will appear to be blue shifted, while the signal from Clock A will appear to be red shifted. Both of these shifts will occur due to the standard Doppler effect.

Let's try another, similar thought experiment.

This time, at the moment when the moving observer passes by the stationary observer, the stationary observer emits a light pulse which travels towards both of the clocks. From the stationary observer's perspective, the light pulse will reach both of the clocks at the same time. But from the moving observer's perspective, the light pulse will reach Clock B before it reaches Clock A.

WHAT THE "STATIONARY" OBSERVER SEES

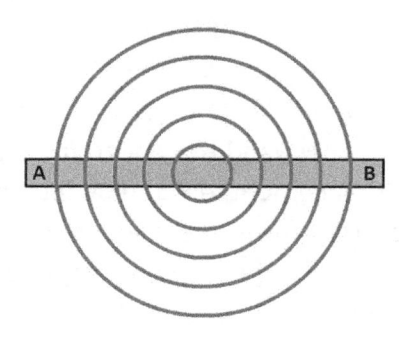

WHAT THE "MOVING" OBSERVER SEES

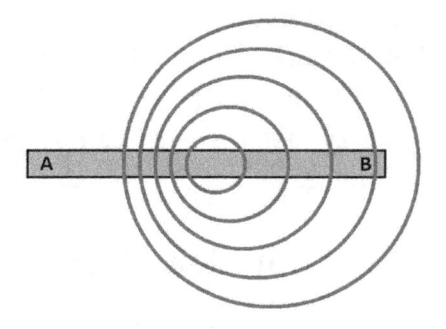

According to SSL theory, both observers are correct, but they are both observing different light. The light that the moving observer sees is not visible to the stationary observer, and the light that the stationary observer sees is not visible to the moving observer. The manifest light that the moving observer sees really does reach Clock B before it reaches Clock A. The manifest light that the stationary observer sees really does reach Clock A and Clock B, at the same time. Rather than a problem of how to define simultaneity, this is merely a problem of confusing different facets of the same superwave.

According to Einstein, perceptions of the simultaneity of events which are separated by vast distances will change, depending on the relative velocity of the observer. In order to arrive at this conclusion, Einstein needed to conflate the true distance with frame-dependent displacement and he also needed to conflate the manifest light with the totality of the light. Under Einstein's model, the totality of the light moves at a singular speed, and this speed must correspond to the frame-dependent displacement, divided by the time elapsed. In order to accommodate different frame-dependent displacements, the time elapsed must also change. By contrast, under SSL theory, only the manifest light which has been selected by the observer moves at a singular speed, and this special speed corresponds to a fixed distance from a particular observer in velocity space.

Under SSL theory it is possible, in theory, to use synchronized light signals in order to determine the simultaneity of events over vast distances, in a way that is objective, and independent from the frame of reference of a particular observer. One merely needs to account for the fact that only manifest light can be seen from any given location in velocity space.

Integrating Local Time with Global Time

If you tighten or loosen a guitar string, then you will alter the rate at which the string oscillates when it is plucked. It is also possible to alter the rate of oscillation of a pendulum through shortening or lengthening it. These are obviously not examples of time dilation. Nobody would be tempted to infer that merely by altering the rates at which these specific objects vibrate, you are also altering the rate at which time itself passes. But if time is defined narrowly and locally then we have no principled reason to reject the notion that time itself speeds up or slows down, as the oscillation of whichever object we designate as a "clock," speeds up or slows down. This is why I claim that it is premature and hazardous to accept any narrow and local definition of time.

Let's compare the sundial with the atomic clock. Both devices measure time, but they employ profoundly different methods. An atomic clock monitors and reports on the oscillation of a single caesium 133 atom, which is less than a nanometer across. A sundial monitors and reports on the rotation and the orbit of Earth around the Sun. The entire Earth-Sun system is about 300 million kilometers across. Therefore, the time that an atomic clock measures is extremely local, while the time that a sundial measures is extremely global. It is very easy to miss this important distinction because the visible objects that we recognize as a sundial, and as an atomic clock, could be comparable in both size and weight. There are also time tracking devices that occupy moderate positions within this continuum, such as an hour glass, or a grandfather clock.

Our original concept of time was derived directly from the motion of the Earth-Sun system. Currently, time is officially defined by the oscillation of a single atom. The definition and measurement of time has therefore transitioned from an extremely global standard to an extremely local standard. The allure of adopting the atomic clock as a standard is largely attributable to the incredible precision of its readings. However, it is important to also bear in mind that extreme precision does not imply superior accuracy.

Einstein believed that time was fundamentally local, as opposed to global, and that time should be defined as merely what a clock measures, as opposed to a relatively objective and abstract dimension. The epistemological orientation of Einstein's understanding of time is one of skepticism and pessimism. In denying the existence of global time, local time becomes the only available substitute. But skepticism with respect to the existence of global time does not warrant a dogmatic acceptance of local time. The idea that time is merely what a clock measures does not imply that a clock cannot also measure something other than time. We still must have methods for identifying broken clocks that depend on variables which incorporate, but also transcend, the clocks themselves. It is always the interaction of the clock with its environment that defines time. Neither the clock, nor the environment, should be considered in isolation. Time is not intrinsic to the local clock, nor is it intrinsic to the global environment. The choice between local and global time is a false dichotomy because time is fundamentally both local and global. Global character is really a property of any object that we recognize as a "clock" and it depends on both the scale of the clock in space and the scale of the clock in velocity space.

If we consider time to be objective, abstract, and global, then we can consider how multiple variables interact, and we can construct a more robust conception of time. A single variable or a single measure may be prone to error or systematic bias, but the overall system should be able to recognize and correct these imbalances. We are not limited to only one method of measuring time or space. We must strive for robust concepts and coherent coordinate systems. A little optimism is all that is required.

Our current concept of simultaneity is inherently pragmatic because we rely on it to coordinate human activities. Intergalactic simultaneity is, by contrast, inherently academic. Suppose that you had an infinite life span and an infinite reservoir of patience. How would you coordinate with an alien from the Andromeda galaxy to meet at a specific planet, in a third galaxy, and at a specific time? On a human scale we have no meaningful interaction with aliens who reside in the Andromeda galaxy. Therefore, we have no practical guidance for how or why we should extend our practical concept of simultaneity in order to incorporate the Andromeda galaxy.

To coordinate with aliens, we would need to know both the distance between us in position space and the distance between us in velocity space. As long as both ourselves and the aliens understand and apply SSL theory then it should be easy to coordinate and meet at the same place and at the same time. Arranging the meeting would be more important than the abstract exercise of synchronizing clocks and it could be achieved without any need to synchronize clocks.

The Need for Diversity in Both Measurement and Definition

Establishing universal standards comes with both great benefits and great costs. In the pursuit of consistency, extreme precision, and perfection, much of the integrity of the basic physical constructs that physics relies on has been sacrificed. It is unfortunate that physics has neglected to establish stronger concordance and corroboration between different methods of measurement. This sacrifice is most evident when it comes to the measurement of time, distance, and velocity.

Consider the following equations: (where **s** represents speed, **d** represents distance, and **t** represents time)

s = d/t

d = (s)(t)

t = d/s

The above equations are perfectly accurate—yet, profoundly inadequate.

Imagine that everything that you knew about speed, distance, and time came only from this set of equations. If that were the case, then you would have no understanding of what any of the corresponding words and symbols mean. Furthermore, you would not be able to differentiate between speed and time because both of these constructs have exactly the same mathematical relationship to distance.

The insularity produced by defining these three concepts solely in terms of one another might not be immediately obvious. But we must have access to other ways of defining and measuring these concepts, otherwise we wouldn't have any way to understand, interpret, or evaluate them. By virtue of being human, we retain the crucial ability to think outside of these equations.

For a concept to be robust and broadly applicable, we must employ a diverse range of methods for defining and measuring it. Otherwise, we may find that we have progressively climbed onto increasingly tenuous limbs that can no longer support the weight of the assumptions that have accumulated.

The idea behind restricting our methods of measurement and our definition for a given quantity is to reduce the manifestation of confusion and bias over the short term. But the latent effect of prematurely reducing this complexity will be to produce more bias and confusion over the long term.

To be fair, measurements and definitions are often restricted for other reasons, as well. Atomic clocks are often used exclusively to measure time because we have no other way to achieve a comparable level of precision. Red shift estimates are used exclusively to measure the velocity of the most distant galaxies because we don't know of any other way to estimate this variable. Using multiple methodologies to estimate the same quantity is sometimes a luxury that we cannot afford, if we wish to reside at the cutting edge of science.

One of the core implications of SSL theory is that relying on the speed of light to define space and time was a colossal mistake. Neither the speed of light nor the atomic clock are sufficiently firm foundations to ground our conceptions of time, space, and velocity. While these errors are deeply entrenched, it is never too late to reverse them.

The Finite and Infinite Manifestations of the Speed of Light

Thus far, I have introduced my Supervelocity and Selection theory of Light, contrasted it with special relativity, and provided a detailed description of it. In what follows, I will attempt to defend the Supervelocity and Selection theory of Light from every conceivable counterargument. I will also be providing a more fulsome review and critique of the currently established version of special relativity. To do this, I will be inspecting both of these theories from a wide array of different angles. Towards the end, I will explore and address the empirical predictions which arise from both of these theories.

To begin with, I will attempt to accomplish something that may seem counterproductive, at first. I am going to lay out a series of arguments in favour of the idea that the speed of light is infinite. I do not really think that the speed of light is infinite, and the SSL theory does not include the proposition that the speed of light is infinite, either. However, my point in making the case for an infinite speed of light is merely to expose the strange, contradictory, and mysterious nature of the behavior of light. What I want the reader to come away with is an awareness of the hazards of oversimplifying this subject matter. Great care must be taken, so as not to rush through the conceptual difficulties that await us.

The Speed of Light as Infinite: Five Lines of Argument

Massless Photons Carry Momentum

The first reason to suspect that the speed of light is infinite is based on the fact that the particles that light is thought to be made up of, known as photons, are considered to be completely massless particles. If you have any doubts about the mass of the photon being zero, then please feel free to peruse the standard model of particle physics and see for yourself. In other words, these particles do not weigh anything, and yet they are still considered to carry momentum.

The momentum that is carried by any object, including an elementary particle, is always a product of its mass, and its velocity. The equation that relates mass, momentum, and velocity takes this form:

p = mv (where **p** stands for momentum, **m** stands for mass, and **v** stands for velocity)

Because a photon does carry some momentum even though it has zero mass, this means that if we enter both the mass, and the momentum, into the formula to determine its speed, then the solution that we will obtain will be a positive number divided by zero, which corresponds to infinity. As the denominator gets smaller, the number gets larger, and as the denominator approaches zero, the number approaches infinity. To be rigorously pedantic, any number is *undefined*, if it is divided by zero, but since we are only considering positive numbers in this case, then we can safely say that the answer must be both positive, and infinite. So, if we trust the laws of physics and the rules of mathematics as they pertain to mass, momentum, and velocity, then we are led to conclude that a massless photon travels at an infinite speed. Because a photon is a particle of light, then this implies that light also travels at an infinite speed.

Independence From the Source

The second reason to conclude that the speed of light is infinite, is the established fact that the speed of light is independent of the speed of its source. This is not true of moving objects, in general. For example, if you throw a baseball while standing on a moving train, the speed of the train will be added to the speed of the baseball, and the baseball will move faster. Because the speed of light is independent of the speed of its source, if a train passes by me at 200 km/ hour, the light that is coming out of the train's headlights will move at the same speed, relative to me, as it would have if the train had been standing still. The velocity of the source cannot add to, or subtract from, the velocity of the light.

This same independence is also a property of infinity. If you add or subtract any number to, or from, infinity, you will always obtain the same result. Infinity plus 200 km/h equals infinity. This is one way to understand how there could be no change. Attempting to add or subtract any speed from the speed of light is equivalent to adding or subtracting a finite number to, or from, infinity. This effort is futile because nothing that you do will have any impact on the result. Hence, this special but finite speed of 299,792 km/s behaves exactly as if it were infinite.

I should address the minor caveat that a similar property of source independence can also be observed with the motion of waves. However, while the velocity of a wave does not depend on the velocity of its source, the velocity of a wave *does* depend on the velocity of its medium. In other words, the velocity of the medium that contains the wave, does add, or subtract to the velocity of a wave. Light, on the other hand, does not move at a fixed speed relative to any known medium. Therefore, it is impossible to say what the velocity of light's medium is. The bottom line is that there is no way to push light in order to increase its velocity and there is no way to pull light in order to reduce its velocity.

Fixed Relationship to the Observer

If light did travel at a finite speed, then it should be the case that if you, the observer, were to move in the same direction as the light, the speed of the light relative to you would decrease by the speed of your motion. And if you were to move in the opposite direction of the light, then the speed of the light relative to you should also increase by the same amount. But the observed reality is that light always travels at the same speed relative to you, no matter how fast, or in which direction you move. One way of looking at this, is to say that it is impossible for an observer to move relative to light. This is why the speed of light cannot be used to determine the velocity of the observer. It is impossible to add or subtract your own speed to the speed of light. Just as in the previous argument, one way to interpret this apparent inability to modify the speed of light through operations like addition, subtraction, multiplication, or division, is to think of its speed as infinite. Thinking of the speed of light as infinite is an effective way of understanding why our finite operations have no influence over it.

Ageless Photons and Instantaneous Travel

It is generally accepted that photons do not age and that they do not experience the flow of time. The theoretical rationale behind this idea is that because photons move at the speed of light, they should also experience infinite time dilation. Given that no time passes for the photon as it travels, from the hypothetical perspective of the photon, no time elapses between its departure and its arrival. Instantaneous travel implies infinite speed. A human observer would theoretically experience instantaneous travel to any destination in the same way if they could also travel at the speed of light.

Physicists generally agree that in some sense it is impossible for any object that has mass to travel faster than the speed of light. Like a rainbow in the distance, it is impossible to ever catch up with a beam of light. Time dilation, length contraction, and inertial mass increase all come into play and conspire to prevent an imaginary spaceship from appearing to travel faster than the speed of light. But this universal law does not apply from the perspective of the passengers on the spaceship. From their perspective, the ship can always continue to accelerate in any direction that they choose, and these subtle relativistic distortions of space and time do not even occur. Because it is always possible to travel a bit further to the west, there is no finite limit to how far west one may travel, in principle. Similarly, a spaceship can always accelerate in any direction that they choose to, without encountering any additional resistance, and so a spaceship should be able to travel at any finite speed that they choose to travel at. The only speed that is permanently out of reach for a spaceship would be infinite speed.

In other words, even without any constraints on speed, one still cannot conceivably travel faster than infinite speed. Similarly, despite the lack of objective constraints that we can point to, it is still considered impossible to travel faster than the speed of light. The speed of light and infinite speed have so much in common that they might as well be the same.

Measuring the Distance to the Rainbow

The rainbow analogy works exceptionally well as a way to think about the velocity of light. We use parallax and binocular vision as methods of estimating distances and both of these methods work according to the same principle. The logic behind parallax and binocular vision is that the less an object appears to change its position relative to other objects, as the observer changes their position, the further away that object must be. Because a rainbow appears to remain in exactly the same location no matter how much the observer moves, this implies that the rainbow is an infinite distance away from the observer. This creates an apparent contradiction when a rainbow appears in a mist that is close to the observer, but it is true nonetheless. By the same token, as an observer moves from location to location in velocity space, light does not appear to move relative to that observer in velocity space. If we apply parallax and binocular vision to velocity space, then we are led to conclude that the speed of light is infinite, as its distance in velocity space appears, by that metric, to be infinite.

Duality and the Speed of Light

I hope to have convinced you that the speed of light is, when viewed from certain angles, apparently infinite. I felt that this was important to do because this is a side of the speed of light which is not often acknowledged. However, the question of whether the speed of light is finite, or infinite is itself a misguided question. While the speed of light is not infinite, it is not finite either. Yet, common sense dictates that a quantity must either be finite or infinite and so it would seem as though there is no third option available. It is strange to think that the choice between finite and infinite could be a false dichotomy, but in this case it clearly is. Light only appears to travel at 299,792 kilometers per second from one isolated perspective at a time, and this is why multiple observers in different inertial frames of reference can never agree on a single, objective speed. The speed of light is finite locally, but infinite globally. The speed of light is relatively finite, but absolutely infinite. The speed of light is manifestly finite, but latently infinite. If you properly integrate and extrapolate from across multiple perspectives, then you will find that the speed of light is all over the place, and this is why light does not behave like the other particles and waves that we know. Speaking of particles and waves, this finite-infinite duality mirrors the wave-particle duality that is already widely recognized. Light is also subjectively particle-like but objectively wavelike.

Aside from the speed of light being incorrectly regarded as finite, singular, objective, and fixed, it is also widely considered to be a universal speed limit. There are significant problems with the idea of light as a cosmic speed limit, as well.

Breaking the 299,792 km/s Speed Limit

Special Relativity and the Universal Speed Limit

The idea that the speed of light is a universal speed limit is what I consider to be a deeply rooted misconception, and I have classified this misconception as Hydra Head #6. It is interesting to note that there is no speed limit mentioned explicitly anywhere in the two core postulates of special relativity. However, the idea of the speed of light as a universal speed limit has still been subsequently imputed into special relativity, as it has been interpreted either as a logical consequence or as an implicit consequence of the two postulates. Conversely, according to SSL theory, the reason that one cannot possibly travel faster than light is analogous to the reason that one cannot outrun their own shadow. A shadow does not really do anything to restrict a runner's speed although it does passively manage to keep up with the pace of the runner.

At the same time, it is both acknowledged that there is a line which cannot be crossed and also that the location of this line cannot be determined. This sort of rule would be very confusing in a game of tennis. This boundary, if it exists, would have to be found somewhere in velocity space. But, for good reason, nobody can come to an agreement on where to draw this boundary in velocity space. This imaginary line also creates a real division between relativity and quantum mechanics, as the principle of locality has been pitted against the principle of realism, in the context of experiments in quantum entanglement. The principle of realism is just the idea that a measurable property of a particle really exists before it is measured. The principle of locality is just the idea that causal influences cannot spread faster than the speed of light. Hence, this same speed limit is also thought to apply to travel, causality, gravity waves, and communication. According to SSL theory, 299,792 km/s is merely an interesting consequence of how matter interacts with light, but according to special relativity, 299,792 km/s has a much more profound, general, and quasi-mystical significance.

According to special relativity, were it possible to do so, successfully travelling faster than the speed of light would entail travelling backwards in time. In this context, the speed of light is acting as a substitute for infinite speed, and for instantaneous travel. Travelling faster than light would necessarily result in arriving at your destination prior to leaving from your point of departure. 299,792 km/s would then become the new infinite speed, as our prior understanding of infinity has been borrowed, and applied to what was formerly a finite quantity.

Obviously, I cannot throw a ball at a window and watch the window break before I release the ball. That would be bizarre. Therefore, it seems as though my influence over the window is limited by the velocity of the ball. It is a matter of common sense that the faster I throw the ball, the sooner it will reach the window, and the sooner the window will break. But it is also a matter of common sense that there is no speed at which I can throw the ball such that the window breaks *prior* to the ball leaving my hand.

As the speed of a ball increases, it approaches an asymptote at infinite speed, and the time elapsed on its journey approaches an asymptote at zero seconds (an asymptote is a dotted line which functions as a graphical representation of infinity). These asymptotes correspond to infinite speed and instantaneous travel, and it would be absurd to contemplate negative speed and negative time on the other side of these asymptotes. But according to special relativity the asymptote is placed at 299,792 km/s rather than at infinity, and this subtle adjustment also creates the impression that one can actually move backwards in time by moving faster than light.

There is a simple but revealing reason for why it is that you can't throw a ball at a window and see the window break before you release the ball. The reason for this is that time always moves forwards by definition, and *everything* moves forwards in time by definition. We wouldn't know how to interpret evidence of a ball moving backwards in time if we were presented with it.

We define *forwards in time* as the direction that time moves in. There are similar conventions for other directions such as right and left or up and down. Asking why up is up and not down or why right is not left is really just a question about the assignment of arbitrary words. But this semantic stability also sets limits on what we can say and what we can think. If you actually did throw a ball and it really did move backwards in time, but in a perfectly continuous manner, then it would just look like the ball had moved towards you and you caught it. You would naturally interpret this event as reversed motion in forwards time, as opposed to forwards motion in reversed time. You wouldn't feel as though you had thrown the ball backwards in time. Instead, you would just experience catching the ball. In other words, by definition, you cannot throw a ball backwards in time because you will always interpret what you experience and observe from a perspective that assumes that time itself is actually moving forwards. A reversal in time must always be interpreted as a reversal in motion. A fast moving ball that is moving towards the left but backwards in time will look exactly the same as a fast moving ball that is moving towards the right but forwards in time. We have no means to discriminate between these two abstract forms of motion, merely by observing them.

I should mention that the second law of thermodynamics can be used to reliably determine the direction of time. However, the relationship between time and entropy is a daunting topic which I have set aside to deal with in my second book. For now, I will just say that the arrow of time is also a product of the dynamics of light. The important relationship between time and the dynamics of light would not be conceivable from within the paradigm of special relativity. But if we can think outside of special relativity, then it is possible to see that the arrow of time is based on the dynamics of light.

According to SSL theory, we cannot observe continuous time travel due to the way that we naturally interpret the flow of time, but this is not due to any finite speed limit. However, if we accept special relativity carte blanche, along with its hyperbolic account of the geometry of spacetime, and of time dilation, then it does make sense to conclude that faster than light travel would also entail backwards in time travel. If we really could observe backwards in time travel, then time loop paradoxes would inevitably result. Because backwards in time travel is forbidden, then according to special relativity, faster than light travel must also be forbidden.

Continuously Accelerating Spaceship

In both popular science and science fiction, the notion that space travel is limited by the speed of light, is normally taken for granted. To travel faster than light, one would require a special warp drive, or some other special technology that can create a loophole to circumvent this limit. But the idea of a warp drive is based on an oversimplified understanding of special relativity, which conflates a mere appearance with a stable, underlying reality.

A light year is the distance that light travels in one year. It seems obvious that if nothing can travel faster than light, and yet it takes light one year to travel one light year, then it must take a minimum of 50 years for anyone to travel a distance of 50 light years, no matter what. Under this interpretation, even travelling a distance of just ten light years within a single lifetime would be difficult, if not impossible. But this is not true, according to special relativity. When special relativity is properly interpreted then it is understood that there is nothing to prevent a person from visiting places that are hundreds or even thousands of light years away, within a relatively short period of time.

If we really wanted to, we could build a gigantic spaceship and load it with hundreds of passengers. So long as we invent an extremely powerful and efficient propulsion system, we could make this ship accelerate in any direction that we want to, at 9.8 metres per second, squared. This rate of acceleration would simulate gravity so that the passengers would feel like they are still on Earth. At this rate of acceleration, the ship would reach the speed of light after roughly one year. After thirty-six years had passed, the ship would be traveling at thirty-six times the speed of light and would have covered a total distance of more than 600 light years. At no point would the engines stop working and the passengers would continue to feel the simulated gravity for the entire trip. Nobody on board the spaceship will see, hear, or feel anything out of the ordinary as the ship reaches superluminal speeds. As far as the passengers are aware, nothing significant is happening.

We can potentially visit any one of millions of planets within our own galaxy without breaking any of the currently recognized laws of physics. However, this project would come with major engineering challenges. For one thing, creating a propulsion system that could accomplish this would be nearly impossible. There would also need to be a way to shield the ship from roaming interstellar objects, which would become very destructive at such high speeds. The barriers to superluminal travel come in the form of mundane engineering challenges but not in the form of the laws of physics.

If you think that this is hard to believe, then this may be because I haven't mentioned the catch, yet. Up until now I have only been describing reality according to the experiences of the passengers on the spaceship. According to special relativity, if the passengers ever decided to return home to Earth, then they would find that Earth had aged considerably, since they had left. The travellers would be much younger than everyone else due to the cumulative and residual effect of time dilation. Upon their return, the space travellers might be interacting with the great-, great, great, great, grandchildren of the people whom they had once known. Meanwhile, from the perspective of someone on Earth, the ship would have never travelled any faster than the speed of light, and the trip would have taken several centuries.

C as a Minimum Speed: C Only Applies to Light

It is important to remember that not only is it frequently claimed that light cannot travel any faster than 299,792 km/s in a vacuum; it is also claimed that light cannot travel any slower than 299,792 km/s in a vacuum. While the velocity of light can be warped and modified in various ways, we need to consider the fact that 299,792 km/s *in a vacuum* is not merely a speed limit, but also a minimum speed. Because we know that nothing other than light obeys this minimum speed, then why should anything other than light obey this maximum speed? Light appears to follow a different set of rules, apart from everything else. It would appear as though light that is travelling in one direction must be moving at 599,585 km/s, relative to light that is moving in the opposite direction. If we can consider an elementary particle such as a photon to be an object, then this seems to imply that one object must travel at 599,585 km/s relative to another object. So, there does appear to be a real contradiction here. Thinking about 299,792 km/s as a minimum speed, rather than as a maximum speed, makes superluminal velocity seem much more plausible.

Distant Galaxies Travel at Superluminal Speeds

Brief Explanation of Red Shift and How to Calculate Speed

We can, arguably, observe distant galaxies to be moving at many times the speed of light. The evidence for this comes from calculating red shift values, which are optical manifestations of the Doppler effect. The Doppler effect occurs whenever waves are traveling at a constant rate relative to a medium, and the source of the waves is moving relative to that medium. The Doppler effect can also be created whenever an observer moves relative to the medium. The waves that are ahead of the moving source are compressed, giving them a higher frequency and shorter wavelength. The waves that are behind the source are stretched out, giving them a lower frequency, and longer wavelength. The faster the source moves through the medium, the more the waves are both compressed, and stretched. Therefore, through measuring the amount of compression or stretching of the waves, and through knowing the speed of the waves relative to the medium, it is possible to determine how fast an object is moving, relative to the medium.

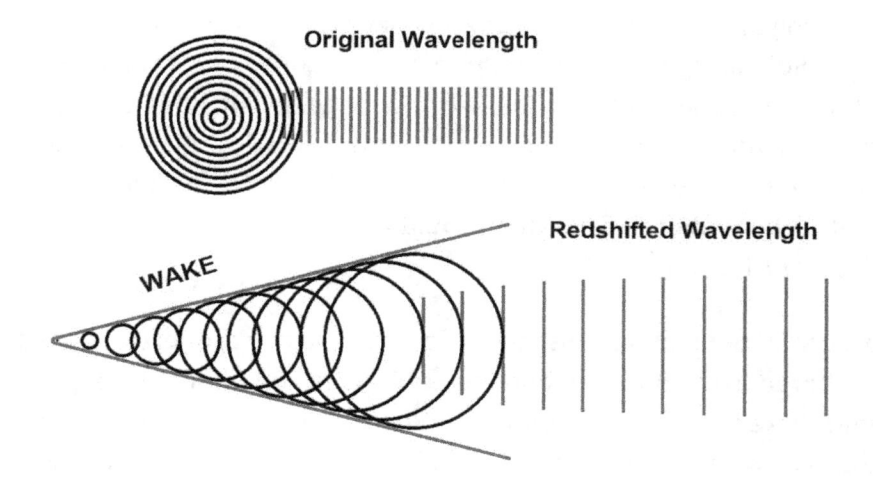

When light waves are compressed, we see the light shift towards the blue, and violet end of the visible spectrum, indicating that an object is moving *towards* the observer. Conversely, when light waves are stretched, we see the light shift towards the red, and infrared end of the visible spectrum, indicating that the light source is moving *away* from the observer.

The Observer-Centred Medium

But what happens when an object actually moves faster than the waves that emanate from it? Motor boats do this all of the time.

The water waves that they produce travel at roughly three km/h, but the boats themselves can travel at more than thirty km/h. If a motor boat is traveling towards you at thirty km/h, then it will reach you before its waves do. That is equivalent to not being able to hear the boat coming with sound, or not being able to see the boat coming with light. A motor boat will also create a wedge-shaped wake behind it. If a motor boat is moving away from you at thirty km/h, its waves can still reach you, because the waves will still move towards you at three km/h, relative to the water, and not relative to the boat.

Thinking about light in the following way is optimal because it helps to avoid making the most common conceptual errors. If the motion of sound waves followed the same strange set of rules that light follows, then sound would work in the following manner: each observer would be totally motionless relative to the air and no matter how they moved, they would not be able to move through the air or feel any wind.

No other medium has this strange and perplexing property of being anchored to the observer and this is why it can be reasonably argued that there is no medium for light. Whether or not there is anything that we would wish to call a "medium" associated with light is a matter of semantics, but if there is any light medium then it would be a very special kind of medium. If water waves behaved as light does, then as each observer moved the entire ocean would also have to move at the same speed and in the same direction. If sound behaved as light does, then the entire atmosphere would somehow need to move at the same speed and in the same direction as each observer.

Light travels through an observer-centred medium. This means that from the perspective of the observer, the source of the light is able to travel relative to the medium, and also relative to the light waves. The light waves travel at a constant speed, but in all directions relative to the medium. The observer cannot travel relative to the medium or relative to the light waves. Instead, as strange as it may sound, both the medium, and the waves, are always pushed or pulled along with the observer so that the observer can remain at the centre, at all times. It is as though this special medium is permanently glued to each and every observer. The medium always shares the same location in velocity space as the observer. This is the right way to think about the observable dynamics of light, even though there are obvious problems with attempting to reconcile the seemingly contradictory observations of multiple observers.

Who is Moving Away from Whom?

From our perspective, it may seem as though it is the other galaxies that are clearly moving away from us, and not the other way around. It is also possible to work out how rapidly distant galaxies are moving away from us, so long as we know how red shifted their light is. Properly applying the mathematics of the Doppler effect leads us to conclude that most of the galaxies that we can see are moving away from us at several times the speed of light.

As a general rule, the galaxies in the universe are moving away from us, as opposed to towards us, and those galaxies that are the furthest away, are also moving away from us at the fastest speed. There are actually many distant galaxies that have an observed degree of red shift which indicates that they are moving away from us at more than ten times the speed of light. The most extreme example that has been found to date, is the distant galaxy GNZ-11, which has a red shift value of 11.1. This value tells us that GNZ-11 is moving away from the Milky Way at 11.1 times the speed of light. In other words, with each passing year, GNZ-11 would appear to be an additional 11.1 light years further away from the Milky Way.

However, even though it is true that distant galaxies are moving away from the Milky Way, it is just as true that the Milky Way is moving away from distant galaxies. If it seems contradictory to assert that the Milky Way is moving relative to multiple galaxies at multiple speeds and in multiple directions at the same time, then this apparent contradiction can easily be resolved by looking at the universe through the lens of velocity space. As I mentioned in chapter two, a map of the known universe would look remarkably similar in either velocity space or position space. Just as the universe has no known centre in position space, it also has no known centre in velocity space. Each galaxy has an objective location, both in position space and in velocity space, but no galaxy is special.

How Could the Light Reach Us?

I have frequently encountered the following question: *If a distant galaxy were moving away from us at faster than the speed of light, then how would its light ever be able to reach us?* This question implicitly assumes that the light would be moving at 299,792 km/s, relative to the distant galaxy which is the source of the light. But this is not how light works, according to the two core postulates of special relativity. From our perspective, the light that we see must appear to move at 299,792 km/s, relative to us only, and not relative to any other galaxy. I am making no attempt to integrate divergent perspectives from multiple inertial frames as it would be premature to construct such a model, at this stage. This is merely a way of accurately modeling and predicting what a particular observer will see from a particular location in velocity space.

If the idea that the light would not be able to travel fast enough to catch up with the observer were true, then this would violate the second postulate of special relativity, in two important ways. Firstly, it would imply that light that is departing from a distant galaxy must be moving at significantly less than the speed of light, relative to the observer who is in the Milky Way's inertial frame. Secondly, it would imply that the light *must* be travelling at 299,792 km/s, relative to the source of the light. The idea that an observer could move faster than the light that emanates from a distant galaxy, such that the light cannot catch up, would thoroughly, and comprehensively contradict the foundations of special relativity.

Simulating Superluminal Speed

For those physicists who will not even entertain the possibility of superluminal speed (even just for the purposes of a thought experiment), I have designed this thought experiment just for you. Even if we assume that superluminal speed is not possible, it must still be possible to simulate superluminal speed in a realistic way.

Let's say that we have set up a row of strobe lights in such a way that they will flash in a timed sequence. Each individual strobe light is stationary, relative to the observer, but collectively they will flash in such a way as to simulate the motion of a rapidly flashing object that would be travelling at 899,377 km/s, or three times the speed of light. Had it been possible for an object to travel at the superluminal speed of 899,377 km/s, while emitting regularly timed flashes at precisely those same locations and times, then that superluminal object would have the same general appearance to an observer.

The only point of this thought experiment is to demonstrate that if an object could travel at 899,377 km/s, while moving away from an observer, and emitting flashes of light, then it really would be possible for the observer to both observe the light that is emitted from that superluminal object and also to use the visual data to calculate the speed of the object and to confirm that it is travelling at 899,377 km/s. The idea that the light emitted by the superluminal object would be unable to catch up to the observer because it is not moving fast enough, is a stubborn misconception. This misconception is often presented in an ad hoc manner, in order to pre-emptively dismiss the idea of considering any possibility of encountering observable evidence for superluminal speeds.

On the other hand, if we were to set up the timing of these flashes in a different way, so as to simulate an object that is moving towards the observer at 899,377 km/s, then something very interesting would happen. In this case, the observer will also see what looks like a flashing object that is moving away from them at 199,862 km/s. But because the observer knows how light works, then they will be able to interpret this information accurately. In other words, the observer would not be able to see such a fast moving object coming, until it has already reached them. Only after the object passes the observer, will they be able to analyze the visual information, and accurately determine how fast the object was moving.

A Distinction Without a Difference: Motion Denial

The established paradigm in astrophysics and cosmology contains a set of assertions about the expansion of the universe which are both mind bending and mundane, at the same time. The main assertion is that galaxies are not really moving apart from one another. While the average distance between galaxies has been increasing over time, this is not due to motion, but due to the expansion of the space that exists between the galaxies. Along the same lines, it is also claimed that observed red shift is not actually the result of the Doppler effect. The "correct" explanation, is that the observed light has been gradually stretched into longer wavelengths because the space that it has been travelling through, has also been stretching over time.

Let's give these two alternative accounts names, in order to distinguish them. I will call the more obvious and intuitive explanation, the "Motion and Doppler" account, and I will call the established explanation, the "Expansion and Stretching" account. What I think is most interesting about these two accounts is that a close inspection of both reveals that we have a distinction without a difference.

Both of these accounts explain an identical set of observations and in a symmetrical way. According to both accounts, the average distance between galaxies has been increasing at roughly 73.3 kilometers per second per megaparsec. According to both accounts, it is possible to calculate the total distance that the light has travelled, as well as the amount of time that has elapsed on its journey, based on the degree of red shift that is observed. The results of these calculations of distance are identical, on both accounts. Also, according to both accounts, the universe is accelerating in its rate of expansion. A useful way of describing the accelerated expansion of the universe is to say that it is *growing* at an exponential rate. The growth of the universe is similar to the growth that is seen with investments and organisms.

There is no conceivable experiment that could be performed, or even imagined, which would be able to distinguish between the Motion and Doppler account and the Expansion and Stretching account. There is no conceivable observation which could tell us which of these two accounts is true. The only real difference between these two accounts is a strong preference for the use of some words in the place of other words.

The common language understanding of the word *motion* is as a change in either position or orientation, over time. The common language understanding of the phrase *relative motion* is as a change in the distances between objects, over time. In other words, if we say that the distance between any two objects is growing, then this necessarily means that the two objects are moving apart from one another. It doesn't matter if the two objects that we are considering are boats, planes, planets, or galaxies. Furthermore, the way that the observed motion is explained or interpreted does not change the fact that it is still motion, and while *absolute motion* is not real, *relative motion* is undeniably real.

The stretching of space has been mathematically modeled and it does work very well as an explanation for the observed red shift. However, these same mathematical models can also be accurately described as models of how the Doppler effect works with light, over vast intergalactic distances. The mathematical model is exactly the same whether we consider it to be a model of gradually stretching light waves over time and space (Expansion and Stretching), or whether we consider it to be a model of gradually shifting frames of reference over time and space (Motion and Doppler). Interestingly, the Motion and Doppler model only works if time dilation is *not* considered in the analysis.

One reason that the Motion and Doppler model is not favoured, is that if the effect of time dilation is considered, along with the Doppler Effect, while interpreting the red shift data, then the results don't make sense, and they don't corroborate with the estimates that are independently derived from luminosity data. You end up with a warped picture of the visible universe that is far too small and oddly centred around the Milky Way. But rather than to discard the time dilation for its distorting effect on the model, it was the Doppler effect that had to be discarded. When push comes to shove, upholding special relativity trumps every other consideration, including the Doppler effect, and the very definition of motion.

The unwritten rules of the discourse stipulate that we can call the change in distance between galaxies whatever we want to call it, but we just cannot call it *motion*. We can also call the apparent stretching of waves, which is ultimately caused by the motion of the observer relative to the source, whatever we want to call it, but we cannot call it the Doppler effect. The big question is, why is it so important to avoid using these special words? In the place of the word *motion*, we have adopted what amounts to a sophisticated euphemism for motion. Occasionally, the term *motion* is allowed, but then the distinction is drawn between *active* and *passive* motion, or between *moving through space* and *moving with space*. In that case, active motion, or moving *through* space would be limited to the speed of light, but passive motion, or moving *with* space, would be unlimited because it is not real motion.

If one theorist claims that there is a difference between *real* motion and *apparent* motion, that is caused by the expansion or contraction of space itself, and another theorist claims that there is no such difference, then the burden of explanation must rest on the person who is adding theoretical complexity by proposing this distinction. Why is it only *some* but not *all* motion that is caused by the changing properties of space itself? Why is this distinction necessary? How is it possible to reliably tell the difference between these two kinds of motion?

The direct and honest answers to the three questions that I posed above are as follows: Motion that appears to exceed the speed of light and violates special relativity, cannot be considered to be real motion, so that special relativity is not shown to be false. Motion that should result in time dilation, according to special relativity, but that does not result in any time dilation, must therefore, not really be motion. This artificial distinction between real motion and a mere change in distance over time is necessary to protect special relativity from being falsified in two separate ways. Special relativity is falsified *either* by the presence of superluminal motion *or* by the absence of time dilation. It is possible to reliably distinguish between these two types of motion because one violates special relativity, while the other doesn't. The rule is that if the motion violates special relativity, then it doesn't really count as motion.

Special relativity has been rescued from refutation through this ad hoc redefinition of motion but at the cost of making special relativity an unfalsifiable theory. Proposing a new type of motion that doesn't "count" as motion is a rather desperate and precarious position to uphold. It seems as though the language of physics and even common sense must bend in order to preserve special relativity. Finding that distant galaxies travel at superluminal speeds and noting the conspicuously absent time dilation in their red shift values should have been the end of special relativity—miraculously, it has survived.

The account of the red shift values of distant galaxies that comes from SSL theory is in many ways identical to the ordinary Motion and Doppler account, but with an interesting difference. For any particular observer, the predictions of SSL theory will be the same as with the Motion and Doppler account, so long as we assume a medium that shares the same location in velocity space as the observer. What SSL theory does in addition to that, is it explains how the observations of different observers, that are moving relative to one another, can still be consistent, and how changes in an observer's velocity will affect their observations of red shift. The basic idea is that the parts of the superwave which are moving faster, relative to the source of light, must also have longer wavelengths, and these longer wavelengths will manifest as red shift. Different observers will observe different degrees of red shift, but merely due to each observer's differing location in velocity space. Contrary to what is asserted by the Expansion and Stretching account, these differences in red shift have nothing to do with the observer's location in ordinary position space, with the amount of time that the light has spent travelling through space, or with the amount of distance covered by the light.

Why Space Itself Is Not Expanding

It is accurate to say that distant galaxies are not really moving in any absolute sense. The motion between galaxies is strictly relative motion. From the hypothetical perspective of the inhabitants of NGC 1300, it is the Milky Way that is red shifted, and it is the Milky Way that is moving rapidly away from NGC 1300. However, it is important to keep in mind that the denial of absolute motion is radically different from the denial of relative motion. In fact, the denial of relative motion is the same as asserting the existence of absolute motion, for all intents and purposes.

The main problem with the idea of expanding space is that it is not clear what it means to say that "space itself," is expanding. Space itself, is not a substance, or an object. Space is not the kind of thing that could shrink, stretch, or expand. Space and time are the systems of coordinates and imaginary grid lines that humans construct. Objects and observers move around within these coordinate systems and also anchor these coordinate systems. If the space around you expanded, how would you know? What would that mean?

There is no way for anyone to tell whether or not the space around them is expanding. Expanding into what? An unspecified space behind the space? Expanding relative to what? An unspecified frame behind the frame? Furthermore, it is not possible for an object to move relative to space itself, at least according to the two core postulates of special relativity. Objects and observers can only move relative to other objects and observers. One can easily move through an ocean, or through the air, but it is not possible to move through empty space, because empty space does not contain the objective reference points that are required to stabilize a frame of reference and to answer the most basic questions about motion. This idea that it is possible to move relative to "space itself," is what I have designated as Hydra Head #5.

When discussing the difference between frame-dependent displacement and true distance, I used the example of taking a trip to the moon. Just as it is impossible to say what the actual length of the path taken by the spaceship is, it is equally impossible to say how much space the spaceship has traversed on its journey. Because the frame of reference is arbitrary, and because empty space by its very nature provides no reference point, it is impossible to say objectively how much space the spaceship has passed through. You might be able to imagine a cubic block of space, just as you could imagine a cubic block of ice, but this abstraction does not correspond to anything in the real world. No such block of empty space actually exists. There is no true answer to the question of how much space there is between Earth and the Moon, even though we do know what the true distance is between Earth and the Moon. Just as it is impossible to move *through* empty space, it is also impossible to move *relative* to empty space. It also follows, that it is impossible for empty space to move relative to an object or an observer. Because space is not capable of moving relative to an object, it also cannot shrink, or grow, relative to an object.

If two galaxies appear to float gracefully apart from one another such that the distance between them is increasing, it would be wrong to say that the galaxies are moving relative to space. It would be equally wrong to say that the galaxies are both stationary, relative to space, and that the space itself is moving. Space itself can neither move nor be stationary, because for space to do either of these things, would imply the existence of an absolute frame of reference. Space would necessarily serve as an absolute frame of reference if it could occupy a defined location in velocity space.

Space Is Not a Fabric, Substance, or Medium

While introducing the concept of velocity space and Galilean relativity, I mentioned that while designing a coordinate system, there are a number of arbitrary decisions that need to be made. We need to select the distance between the grid lines, the position of the grid lines, and the orientation of the grid lines. So far, there is no problem. But the fourth decision that we must make, is to select a location zero in velocity space for our grid lines to occupy. This is a huge problem because it means that the analogy of a medium, a fabric, or a substance does not fit.

One thing that all known fabrics, substances, and mediums have in common, is that they have a predefined location in velocity space. For example, a fish can swim relative to the water, and a bird can fly relative to the air because the water and the air each occupy defined locations in velocity space. As bread rises, the particles in the bread also move, relative to one another. As a balloon expands, the material stretches, which implies motion. A typical medium would have distinguishable parts, and those parts would be capable of relative motion. One cannot move through space, or relative to space, and space cannot move relative to an observer, either.

In summary, expansion necessarily implies a defined volume at *time one*, and a defined volume at *time two*, it also necessarily implies a defined set of locations in space for the object or medium that is expanding and a defined location in velocity space for the centre of the object or the medium that is expanding. Expansion also necessarily implies that one part of the object or location within the medium is moving, relative to another part, or location. Therefore, the space that is said to be "expanding," would need to have both a defined position, and a defined location in velocity space. If the coordinate system has both a defined position and velocity, then it becomes an absolute frame of reference. Therefore, it is not possible for space itself to expand, contract, stretch, or move without violating the two postulates of special relativity. According to the two postulates of special relativity, all locations in velocity space must be on an equal footing. But if space itself were to occupy a particular location in velocity space, then this particular location in velocity space would be very special. Therefore, space itself cannot occupy any particular location in velocity space, just as velocity space cannot occupy any particular location in space itself.

The Minkowski Spacetime Manifold

According to special relativity, every object in the universe is always moving at 299,792 km/s, through spacetime. Spacetime is construed as a hybrid manifold which integrates three dimensions of space with time. However, even though the speed of all observers and all objects is purported to be held constant in spacetime, it is still possible to vary one's direction of motion through spacetime. As a result of adjusting the direction of motion, the claim is that the faster one moves through space, the slower one moves through time, and vice versa. If this concept of spacetime is difficult to understand, then rest assured that you are in good company. In my experience, the more effort that is put into understanding spacetime, the more elusive it becomes. Fortunately, I have eventually arrived at a clear understanding of what makes spacetime so difficult to grasp. Spacetime is confounding because ordinary concepts such as *time* and *speed* change their meanings significantly when they are presented within this paradigm. *Time* and *speed* are employed metaphorically but not literally within the spacetime model. The notion of the direction of motion through spacetime, is also highly abstract, and metaphorical. Similarly, a *distance* in velocity space is not really a distance, but a speed.

There are obvious epistemological and logical questions that arise from the claim that the faster one moves through space, the slower one moves through time. How would I determine how fast I am traveling through time? We have already established that there is really no such thing as how fast any observer or object is moving through space itself, but if we take the idea of moving through time at a certain speed, literally, then we will also arrive at logical contradictions. If speed is defined as distance over time (as it normally is) and if time is defined as distance over speed (as it often is), then how can we say anything definitive about the "speed" at which anything moves through "time," without radically altering the definitions of both speed *and* time? Let's just say that my speed through time is 100 km/s. 100 km/s is a real speed, but it isn't really a speed through time. On the other hand, let's say that time for me is passing at ten percent of the rate that time is passing for my environment. In this case, we have a measure of the rate at which I am advancing through time, but this quantity is not really a speed, according to the standard definition, because it is not measured in units of distance over time. It is also not clear which of the two times should count as the "time" which an observer is passing through. Is "time" a reference to the time of a specific observer or does it refer to the time of the observer's environment?

If it is possible to get oneself further into the future while aging by the same amount, then what does this imply about the rate at which one must have travelled through time? Would this rate have been faster or slower? The fact of being able to experience a wider span of time, of being able to see further into the future, and of having covered more "time" in the same amount of "time," would seem to imply that one is travelling through time at a *faster* rate. I would suggest that speeding up is the most natural interpretation, but this is not the interpretation that has been adopted and implied by the Minkowski spacetime manifold. According to the standard interpretation of spacetime, aging more slowly relative to one's environment is considered to signify that one is advancing *more slowly* through time. The claim that an observer moves more slowly through time, as they move more rapidly through space, is just a restatement of the idea of time dilation, which I will be discussing at length, in the next chapter. Because the spacetime model is founded on the assumption that time dilation from inertial motion is real, then if time dilation from inertial motion is not real, then the entire framework of spacetime must also be misconceived.

The situation that this poetically postulated spacetime model is trying to describe, is really one where a particular person or object would theoretically experience relatively fewer changes than another person or object would, over the same *time* period. In essence, this means experiencing less motion in general over the same *time* period. Moving less in the same amount of time is very different from covering less "time," in the same amount of time. If it is a mistake to conflate distance with time in this way, then it is a deep, and profound mistake.

What all of this implies is that there are two kinds of motion contemplated by special relativity. The first kind of motion is inertial motion in a straight line (which corresponds to a defined location in velocity space) and the second kind of motion is directionless motion (motion "through time"). According to special relativity, the faster one moves in a straight line, the slower one's directionless motion becomes, and vice versa. Directionless motion is called *motion through time,* and motion in a specific direction is called *motion through space*. For example, if a cannonball clock were fired into space, then its cannonball-like motion would be its motion through space, and its clock-like motion, including its ticking, and everything else that the cannonball clock does which contributes to its ticking, would be its motion through time. Motion "through time" just means all motion that doesn't qualify as inertial motion. Motion "through time" could also be a physical rotation, as in the rotation of the gears in a clock, or an oscillation, as in the vibration of a spring. Special relativity merely says that the more cannonball-like motion that the cannonball clock engages in, the less clock-like motion it engages in, and vice versa.

My main point about spacetime is that the spacetime model has actually separated motion into two antagonistic types of motion. Spacetime does not really contemplate *time* and *space,* so much as it contemplates *inertial motion* and *directionless motion.* Just as inertial motion is relative, so is directionless motion, because in order to measure time we must rely on directionless motion. The model of spacetime also presumes that everyone and everything is allotted a certain quantity of total motion per unit of time, such that the more inertial motion that an object engages in over a given period of time, the less directionless motion it will engage in over that same period of time, and vice versa. Breaking motion into inertial motion and directionless motion gives us a much clearer way of understanding what the Minkowski spacetime manifold says about the structure of reality and the interplay of space and time. This also makes it clear that the "speed" of both types of motion relies on an additional measure of time which is extrinsic to the model. Time manifests as directionless motion from within the model, but it also implicitly exists somewhere outside of the model in an absolute form. This dual understanding of time is inherently confusing, as it poses a formidable challenge to anyone who seeks to pin down a clear, singular, and coherent definition of time.

Spacetime as Represented in Velocity Space

If we apply the rules of special relativity to velocity space, then our stable, and Euclidean map will become dynamic and hyperbolic. If spacetime were represented in velocity space, then the aforementioned "light bubble" (the distance of 299,792 km/s from the observer in all directions in velocity space) would become the external border of the map. All distances in velocity space would then gradually become distorted and flattened as they approached the bubble-shaped border of the map. Velocity space would now have a finite volume. The entire map would also look different to each observer. As an observer moves within this space, the objects would also change their relative locations. The distances between the objects themselves would also change. The map of velocity space would wind up becoming a profoundly subjective and dynamic map. In fact, arguably, this new map would not really be a "map" at all, because a real map should be relatively objective, and static. The locations of the landmarks on a real map should not shuffle around and reorganize themselves, depending on where the observer is standing. Distances and angles should not change, depending on where the observer is standing. The whole point of having a map is to transcend your limited perspective and to attain a more universal and objective understanding of the world around you.

A functional map should be able to give you a sense of where you are and how to get to where you want to go. A map is not intended to tell you only how things look from where you are standing. You don't even require a map to tell you how things look from where you are standing. This conspicuous redundancy is what I find frustrating about special relativity. Special relativity seems more like a system of awkward and ad hoc explanations, than a useful guide for navigation. While often praised for its coherence and its elegance, special relativity is anything but coherent, and elegant. It is now time to dig deeper into special relativity.

Thoroughly Examining Special Relativity

The Three Corrective Distortions of Special Relativity

Einstein's Light Clock Thought Experiment

In addition to the two postulates of special relativity, the light clock thought experiment is the core model that defines special relativity. The light clock thought experiment allows us to ascertain what special relativity proposes and predicts so that we can derive, validate, and interpret its equations. It is only through deriving the Lorentz equations yourself, that you can fully appreciate both the assumptions that are incorporated into them, and their limitations.

We begin with the concept of a light clock, which is based on the logical proposition that if light maintains a constant speed, then you should be able to use the speed of light to keep an accurate record of the passage of time. A light clock would be a special device which reflects a laser pulse back and forth in a vacuum between two parallel mirrors. Each time the light strikes a mirror, one unit of time is recorded as having passed.

There are actually two versions of the light clock thought experiment. The standard light clock thought experiment is known as the *transverse* light clock, and this thought experiment is used in order to make the case for time dilation. The less common version of the light clock thought experiment is known as the *longitudinal* light clock, and this thought experiment is used to make the case for both time dilation, and length contraction.

In the light clock thought experiment, the novel concept of time dilation is introduced as a means of explaining how it is possible that the moving observer can see the light clock that is in their inertial frame of reference tick at the normal rate, even though the observer in the stationary frame sees the same light clock tick, at a slower rate. The light clock thought experiment makes the contradiction that is implicit in a constant but relative speed of light clearer, and it also proposes that the solution to this contradiction is that the rate of the passage of time must change between the two observers, in order to bring their otherwise inconsistent observations back into alignment. Time itself must bend in order to maintain a constant speed of light, according to special relativity.

In reality, a light clock has never been created, and it could never be created, for a variety of reasons. We would need an impossibly perfect laser that emits perfectly coherent and monochromatic light in a beam which never widens. We would also need impossibly perfect mirrors that are perfectly flat, and which reflect all of the light. We would also need a way to observe and record each impact of the laser on the mirror, without diverting the beam, or draining its energy. The laws of physics conspire to preclude the possibility of ever developing a functional light clock, and so the light clock will never be anything more than a thought experiment.

But let's say that we do have a working light clock and that it will be placed on a space ship and that the spaceship will appear to be traveling at half the speed of light, relative to an observer on Earth. The mirrors inside this light clock are positioned to face each other at an angle that is perpendicular to the direction of the relative motion of the spaceship.

In the light clock thought experiment, the observed direction of the light inside the light clock depends on both the speed, and the direction of the ship, relative to the observer. As the relative speed of the ship approaches the speed of light, the apparent direction of the light inside the light clock also approaches the apparent direction of the ship, from the observer's perspective. If the spaceship were moving at the speed of light, then from the space traveler's perspective, the light would be moving perpendicular to the direction of motion but from the Earth based observer's perspective, the light would be moving parallel to the direction of motion. There is a ninety-degree difference between these two observations. This ninety-degree difference in direction cannot be accounted for by either time dilation, or length contraction, or both, but I digress.

Time Dilation

Let's say that the spaceship with the light clock on board is travelling at half the speed of light, relative to Earth. From the frame of reference of the observer on the spaceship, the light in the light clock would be bouncing directly back and forth with a change in direction of 180 degrees, after each reflection. But from the perspective of someone on Earth, the light would appear to be reflecting back and forth at 120 degree angles. Because the light appears to be following a 15.47 percent longer path in between reflections, from the perspective of someone on Earth, then special relativity has it that there must also be about 15.47 percent more time in between each of those reflections, from the perspective of an observer on Earth.

If you extend this logic, then you can conclude that for every six minutes that appears to pass for a traveller on the spaceship, roughly seven minutes would have passed for the observer on Earth. Relatively less time passes on the spaceship, from the perspective of the observer on Earth. Time on the spaceship has therefore slowed down, relative to time for the observer on Earth. Because speed is defined as distance divided by time, speed can be modified if either time or distance is modified. One possible way to maintain a universal, relative speed of light is to allow the flow of time to be distorted. But when we say that time slows down on the spaceship, this really only means that motion in general has slowed down, that the light clock has slowed down, and that the person on the spaceship is aging more slowly. The slowing of time that is expected whenever an object moves at a high velocity is known variously as either kinematic time dilation, relativistic time dilation, inertial time dilation, or just time dilation. I will be referring to it as *inertial time dilation* because I think that this provides the clearest expression of the concept, and also to distinguish it from gravitational time dilation. The concept of time dilation is easy to imagine if you think about playing a video recording in slow motion, but this slowing down would only be visible to an observer who is observing from a different location in velocity space.

Although it is associated with Albert Einstein, Henri Poincaré, Hendrik Lorentz, and Hermann Minkowski, the Lorentz transformation equation was actually developed by Woldemar Voigt, in 1887. This special equation is used to determine either the degree of inertial time dilation, or the degree of length contraction, or the degree of inertial mass increase, given the relative velocity of the object. What the light clock thought experiment is intended to illustrate, is that both the total distance travelled by the beam of light, and the angle of the beam of light, will appear to be different when viewed from different locations in velocity space. One hypothetical effect of this is that the rate of ticking of the light clock will appear to slow down as the relative motion of the light clock speeds up. The Lorentz transformation, which describes the relationship between velocity and inertial time dilation, is depicted below.

$$\gamma = \frac{1}{\sqrt{1-\left(\frac{v}{c}\right)^2}}$$

Apparent Time Dilation γ

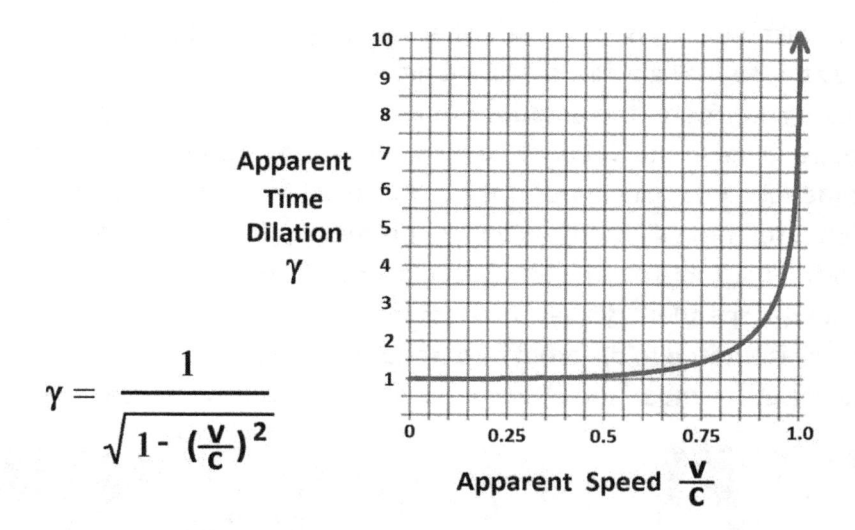

Apparent Speed $\frac{v}{c}$

This light clock thought experiment works well, but only if the light clock is oriented in just the right way. If instead, we were to rotate the light clock so that the light is now reflecting back and forth in the same direction as the motion of the spaceship, then we will see different results. After running the same calculations again, we will find that instead of an increase of 15.47 percent of the total distance travelled by the light, we get an increase of 33.33 percent this time. This makes no sense! Why would the amount of time dilation change dramatically, depending on how the light clock is oriented?

Not only do the results of the light clock thought experiment change depending on the orientation of the mirrors, but you will also get results that don't make sense for other reasons. A triangular-shaped or circular-shaped light clock would also yield different results. The reason for these differences is that the light takes less time to travel from the front mirror to the back mirror, from the Earth's frame, than it does to travel from the back mirror to the front mirror. I mentioned previously that this modification on the standard light clock thought experiment is known as the *longitudinal* light clock. The fact that the light takes a different amount of time depending on its direction, is tied to the different sizes of its frame-dependent

120   Geoffrey Stone

displacement. This also means that the speed of the light is different, relative to the light clock, depending on which direction it is traveling in. Furthermore, the amount of inertial time dilation also changes, depending on the direction of the light. Not only does the degree of time dilation change, but it is better described as *time contraction* (inverted time dilation), as the light is moving from the front mirror to the back mirror. The images below depict how the path and the frame-dependent displacement of the light both change as the light clock is rotated to different orientations. In these diagrams, a longer light path is understood to be equivalent to a longer interval of time.

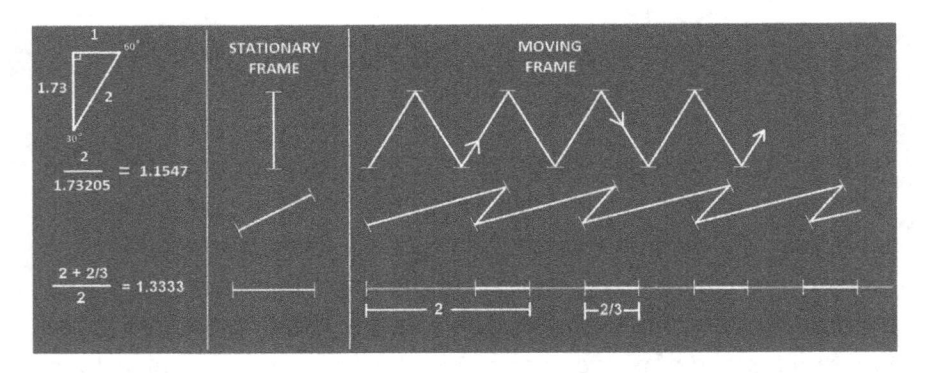

When inertial time dilation is calculated, this is done as if inertial time dilation were strictly a macro-property of the behavior of light, which cannot be analyzed in terms of its individual components. The calculated inertial time dilation which results from the Lorentz transformation is only an average taken by combining the trip from the back mirror to the front mirror with the time taken in the opposite direction. This averaging avoids and obscures the fact that the time dilation factor should actually vary, depending on whether the light is moving towards, or away from the direction of relative motion. With the longitudinal light clock, if the spaceship is moving at half the speed of light, then the forward-moving time dilation is a factor of exactly two, but there is also a backward-moving time contraction, of two thirds. This is a very important problem to remember, and as I will explain later, this is a clue as to why inertial time dilation is only effective at resolving the big contradictions from within the narrow confines of the light clock thought experiment. Outside of this fragile and contrived light clock thought experiment, time dilation is useless. Just as a magician can no longer pull a rabbit out of a hat if he was prevented from depositing the rabbit into the hat initially, inertial time dilation, and length contraction cannot make sense out the behaviour of light, if we tinker with the fragile conditions that constitute the light clock thought experiment.

If we look at the light clock thought experiment through the lens of SSL theory, then we can find a simpler way to resolve the apparent discrepancy. According to SSL theory, the stationary observer would see the moving light clock tick at the normal rate, even though they would see the light bouncing back and forth between the mirrors, at a longer time interval. The explanation for the divergence between the ticking of the clock and the reflecting of the light, is that the light that the stationary observer sees is not the same light that is causing the clock to tick. The stationary observer sees a slower part of the supervelocity, and the clock mechanism responds to a faster part of the supervelocity. Hence, both observers would see the light clock tick at exactly the same rate.

Length Contraction

Einstein arrived at a brilliant solution to the problem of getting a different average value when rotating the light clock, through applying the concept of length contraction. When the light clock is oriented so that the beam of light aligns with the direction of motion (the longitudinal light clock), the clock would appear to be ticking even slower, all else being equal, but it won't actually appear to tick any slower, so long as the length of the light clock also appears to have shortened, such that the distance between the mirrors (the true distance) has also decreased. According to special relativity, the mirrors of the longitudinal light clock must also appear to get closer together, and this explains why we get the wrong answer as we calculate time dilation in this special case. This shortening in the distance that needs to be covered must be exactly the right amount, in order to compensate for the slower speed of the light. This would mean that a circular object would appear to be increasingly elliptical as it approaches the speed of light, as it would appear to be squished, or flattened, in the direction of motion. But of course, from the perspective of the observer on a moving spaceship, there would not have been any noticeable inertial time dilation, or length contraction, with respect to their vessel.

Because speed is defined as distance divided by time, one can either influence speed through distorting time or through distorting distance. Time dilation is a distortion of time, and length contraction is a distortion in distance or space. Both of these distortions are widely considered to be necessary, in order to ensure that light maintains a constant relative speed with respect to all observers, in all inertial frames of reference.

Inertial Mass Increase

Given that momentum is a product of velocity multiplied by mass, and given that kinetic energy, and momentum, are conserved quantities, then what is happening to the mass of the spaceship as it travels at half the speed of light, relative to the earth? From the perspective of an observer on Earth, the spaceship would appear to be traveling much slower than it should be, due to time dilation. In other words, the traveller appears to be traveling more slowly than they actually travel, and they appear to be covering less distance than they actually cover, but yet the total momentum of the spaceship must somehow remain exactly the same, otherwise this would violate conservation of momentum. If we assume that the speed of an object is capped at 299,792 km/s, where there is also an asymptote on the graph, then it begins to look as though the rate of increase in momentum must be outpacing the rate of increase in speed, especially as we approach the asymptote. It is on the basis of these tenuous assumptions that an increase in mass has been hypothesised.

Given that the speed of the spaceship appears from the perspective of someone on Earth to have been reduced by 13.40 percent, due to time dilation, something has to give, in order to conserve momentum. In order to balance the momentum equation, the apparent inertial mass of the spaceship must increase by 15.47 percent. This means that, if the spaceship were to crash into a planet, it would still strike the planet with as much force as if it had been travelling 15.47% faster than it appeared to be. From the traveler's perspective on the spaceship, everything on the spaceship would still have exactly the same mass. However, from the perspective of an observer on Earth, some of the energy from the rocket fuel has been converted into extra mass, instead of extra speed.

Inertial Mass vs. Gravitational Mass

For various reasons, mass can also be separated into its inertial, and gravitational relationships. Gravitational mass is the aspect of mass that interacts with the force of gravity. More massive objects exert a stronger pull on nearby objects through gravity, and more massive objects also weigh more under the same force of gravity. On the other hand, inertial mass is the resistance that massive objects exert against any attempt to adjust their location in velocity space. Interestingly, according to special relativity, the relativistic effects of traveling at close-to-light speed, only increase the apparent inertial mass of the fast moving object. The high speed does not have any effect on the gravitational mass of the object. What this implies is that it is not the mass of the object that changes per se, but only the inertial facet of the mass.

The most cogent explanation for why these two types of mass should vary independently, is the fact that the mass of the object does not actually change, it only appears to change from a narrow perspective. So-called "relativistic mass" is just a useful fiction. So long as a spaceship is in the vacuum of space, it is always equally easy, and equally difficult to accelerate it in any chosen direction. The inertial resistance to acceleration does not change depending on the past history of acceleration, and it does not change depending on where the spaceship is in velocity space.

There are very good reasons to believe that relative velocity does not have any effect on gravitational mass, even though some physicists claim that it does. If gravitational mass was actually variable in this way, then the motion of an observer would have vast, and profound effects on the world around them, which cannot possibly be real. If one was to board a spaceship and travel at half the speed of light, relative to the Sun, then, from the space traveller's perspective, the gravitational mass of the Sun would need to increase by 15.47 percent. This increase in gravitational mass would severely disrupt the orbits of all of the planets in the solar system, and it would cause total chaos. But this disruption could only occur from the perspective of the observer. This outcome would be absurd, but intuitively it is easy to recognize that something which is fixed and objective, like gravitational mass, could not possibly depend on the subjective perspective of an observer.

Common Themes in Special Relativity

Because the apparent speed of light cannot change, then it is assumed that time itself must be flexible, in order to compensate for changes in the apparent length of the path of the light. Because the apparent speed of light cannot change, then it is assumed that space itself must be flexible, in order to compensate for changes in the apparent length of the path of the light. Because the apparent speed of an object cannot exceed the speed of light, then it is assumed that inertial mass must be flexible, in order to compensate for the apparent reduction of speed, and to conserve momentum. All three of these distortions (inertial time dilation, length contraction, and inertial mass increase) are smooth and gradual and all of them are limited only to the subjective perspective of an observer. Time, space, and mass are treated as mere matters of perspective, under special relativity. The purpose of all three of these concepts is to function as a bandage, in order to repair what would otherwise result in the violation of an important physical principle. But these three solutions create far more problems than they solve.

The Hidden Complexity and Contradiction in Special Relativity

Does the Direction of Motion Matter?

There is one glaring flaw that is built into Einstein's light clock thought experiment, which I have already mentioned. In the standard text book version of the light clock thought experiment, the light beam is always reflected perpendicular to the direction of motion. Whereas in the longitudinal light clock thought experiment, the light is reflected in the same direction as the motion of the spaceship. From the hypothetical perspective of the observer on Earth, the light does not move inside of the longitudinal light clock, at the same speed, in both directions. In fact, it moves much slower, relative to the spaceship, while it is traveling in the forward direction, and much faster, relative to the spaceship, while it is travelling in the backward direction. This would mean that the effects on time for the spaceship would also toggle back and forth between time dilation and time contraction, as the light bounces between the mirrors. This conclusion is absurd, but the absurdity can be avoided if we consider only the total time, or the average time, that the light requires for a round trip.

If it was only the one-way speed of light that was to be considered, then other absurd conclusions would follow. For example, if while riding on a spaceship, a traveller were to shoot a laser beam in front of them in the same direction as their motion, relative to Earth, then an observer who is stationary with respect to Earth, would conclude that space traveller's time is dilating, so that the observer on the spaceship can see the same beam moving faster. But if the space traveller were to aim this same laser beam behind them, then the opposite effect would occur. They would need to appear to experience time contraction, in order to see the same beam moving slowly enough to match what the other observer sees. This result is also absurd, because it should not matter in which direction one aims their laser beam. The engagement of the laser beam is not really what causes time dilation or time contraction (which should be impossible). It is only the speed of relative motion of the spaceship that is thought to cause the distortion in time. Furthermore, nothing is preventing the traveller from shining a light both ahead and behind them at the same time, and this would lead to a blatant paradox. What makes this paradox especially problematic, is the fact that this is what normally happens, when light is emitted. Light normally travels in all directions simultaneously. But the main point that I want to highlight here, is that the one-way speed of light is considered to be both irrelevant, and deeply problematic, with respect to calculating time dilation in special relativity.

Thinking Outside of the Light Clocks

In this simple thought experiment, there is an observer on the ground, a car driving eastward at eighty km/h, and a bird flying eastward at one hundred km/h. But this bird is very special because it follows the same rule that light follows. Let's assume that both the observer in the car and the observer on the ground agree on precisely how fast the car is moving, relative to the ground. Perhaps, the observer on the ground has a radar gun and the observer in the car has an odometer, and they both measure eighty km/h. From the perspective of the observer on the ground, the bird would be moving at one hundred km/h eastward, relative to the ground, the car would be moving at eighty km/h eastward, relative to the ground, and the bird would be moving at twenty km/h eastward, relative to the car.

But Where is the Rainbow?

From the perspective of the observer in the car, the bird would be moving at 180 km/h eastward, relative to the ground, the car would be moving at eighty km/h eastward, relative to the ground, and the bird would be moving at one hundred km/h eastward, relative to the car.

These relationships are depicted in the velocity space maps below. The person on the ground is represented by a circle, the person in the car is represented by a square, the bird is represented by a triangle, and the role of the observer is circled.

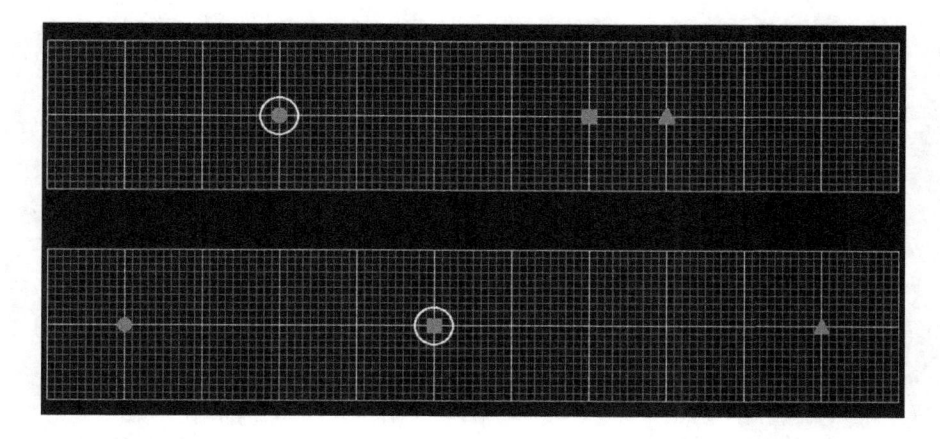

As shown in the two velocity maps, the accounts of different observers are in conflict because the observer in the car sees the bird moving at one hundred km/h, relative to the car, while the observer on the ground sees the bird moving at only twenty km/h, relative to the car. Furthermore, the observer on the ground sees the bird moving at one hundred km/h relative to the ground, while the observer in the car sees the bird moving at 180 km/h, relative to the ground. Interestingly, the difference between divergent accounts always equals eighty km/h because that is the difference in velocity between the two observers.

Obviously, the bird cannot be both moving at one hundred km/h and also at 180 km/h, relative to the ground. The bird also cannot be both moving at twenty km/h and also at one hundred km/h, relative to the car. The big question is, how is it possible to resolve these discrepancies? It turns out that special relativity cannot supply us with an answer to this question. The reason that special relativity cannot supply us with an answer to this question is that we have been contemplating the one-way speed of light with this example. Special relativity never really contemplates the one-way speed of light, even though it is normally presented as though it does.

Mapping the Subjective onto the Objective

Psychophysics is the branch of science which studies and maps out the relationships between objective and subjective variables and it is also a field that I have some academic training in. There are two psychophysical measures which have become widely used in meteorology. Humidex and wind chill are both methods of reconciling how weather conditions feel, with how they really are. The humidex value considers objective measures of temperature and humidity, in order to give you an idea of how hot the weather should feel. Wind chill considers objective temperature and average wind speed, in order to arrive at a value that represents how cold the weather feels. With each of these measures, there is an explicit standard for calibration. The subjective temperature is always equal to the objective temperature, so long as the humidity is zero, and the wind is absent.

With the velocity of light, we can also witness a divergence between objective and subjective values. But in this case, it is not clear how to calibrate or relate these two values to each other. Under what special set of circumstances is the subjective velocity of light equal to the objective velocity of light?

The rainbow analogy comes in handy, as we attempt to map the observations of light's behaviour onto a deeper framework. Under which special set of circumstances is the objective position of a rainbow equal to the subjective position of a rainbow? In the case of the rainbow, the answer is that there is no way to map the subjective appearance of the rainbow onto the real rainbow. There is no correspondence between where the rainbow *appears* to be and where the rainbow really *is*.

Where is the Rainbow?

There is a subtle but illuminating difference between the rainbow that an observer may see accompanying a thunderstorm and the rainbow that can be created with the help of a prism. If a prism is used to project a rainbow onto a wall, then there really would be a rainbow located in a definite position on that wall. An observer could then point to this rainbow and correctly identify where the rainbow is.

But in the case of a naturally occurring rainbow, the only places where the rainbow really exists would either be in an image that is formed on the back of an observer's retina, or in an image that is taken of the rainbow with a camera. In the absence of retinas and cameras, there is no rainbow to be found anywhere. An image must first be created somewhere, in order for the rainbow to take shape. Once the rainbow does take shape, it only takes shape in the location where the image itself is created. It really is true that the rainbow does not exist, as such, unless or until someone looks at it.

A real rainbow does not ever take shape in the sky. While one can still point in the direction of the rainbow that one sees in the sky, the sky is not actually where the rainbow exists. Any two observers who both point to the "same" rainbow, will actually be pointing to two different locations in the sky. The rainbow-shaped object that each of them sees in the sky is merely an illusion. For proof of the illusory nature of the rainbow, one only needs to attempt to locate the point where the rainbow appears to touch the ground. Not only will it be impossible to find this special location, but there will also be no pot of gold, and no leprechaun in sight. Hence, there is no way to map the subjective experience of the location of the rainbow onto any particular objective location.

Similarly, with the velocity of light, there is no way to map the subjective experience of the velocity of light onto the objective velocity of light. The reason that this mapping cannot be accomplished in terms of the velocity of light, is due to the total absence of any objective velocity. One cannot map the subjective perception onto the objective reality when there is no corresponding objective reality.

A rainbow is merely an image with no corresponding object. As an image, a rainbow is represented only as a circular spectrum of colour, in two dimensions, with red on the outside, and violet on the inside. But it is also very easy to imagine a rainbow as a real object because rainbows are commonly represented as real objects in cartoons, in fantasy, and in Skittles™ commercials. If rainbows were real objects in real life, then our interactions with them would be radically different. The experience of interacting with such a rainbow would be similar to exploring the Gateway Arch monument in St. Louis, Missouri, if that monument were more evenly curved, and if it had been given a bright and colourful new paint job. Here is a list of some of the key observations that one would expect to make while inspecting a rainbow, if a rainbow were a real, three-dimensional object.

1. As you approach the rainbow, it should loom larger over the horizon. It will also appear to shrink as you walk away from it.

2. It would be possible to view the rainbow from underneath. From underneath, the rainbow should look vanishingly thin, due to its ribbon-like shape. It would also be possible to cross to the other side of the rainbow and view it from behind.

3. One would be able to walk towards one side of the rainbow and find the specific location where the rainbow meets the ground.

4. One could view the rainbow from an oblique angle, such that the arc would no longer look perfectly round and circular but relatively tall and skinny and with a sharply curved peak, at the top. The rainbow would then manifest as a vertical ellipse. The side of the rainbow that is closer would also look wider than the side that is further away.

5. It would be possible to use depth cues from either binocular vision or parallax, in order to get a sense of how far away, and how large the rainbow actually is.

By contrast, while observing a real rainbow, its apparent size will remain fixed as you walk towards or away from it. In effect, this means that a rainbow appears to grow as you walk away and shrink as you approach it. It is not possible to look at a real rainbow from underneath, from behind, or from an oblique angle. One side of a rainbow can never appear to be closer than the other side. Finally, parallax and binocular vision will not provide us with any finite details about how large or how far away a real rainbow is.

Try to imagine explaining the strange behavior of a rainbow to someone who thinks that a rainbow is a real object, like the Gateway Arch, but without informing them that a rainbow is merely an image formed on their retina, and nothing more. I imagine that the behavior of a rainbow would be very difficult to explain without dispelling the central misconception of the rainbow as a real, 3-dimensional object. The strange behavior of the rainbow could easily be described and modeled with a series of equations, but such equations would not serve as the basis for any genuine explanation.

Special relativity regards the speed of light as a stable fixture in the physical landscape—much like the Gateway Arch is, in St Louis. By contrast, the Supervelocity and Selection theory of Light explicitly recognizes the manifest speed of light as nothing more than an appearance, much like a real rainbow. To ask the question, "what is the speed of light?" reveals an implicit, and naïve misconception. The false assumption behind that question is just as salient as the false assumption behind analogous questions about rainbows such as: How tall is this rainbow? How far away is this rainbow? Where does this rainbow touch the ground? What is the angle of orientation of this rainbow? Et cetera.

I also want to be crystal clear about how the rainbow analogy fits into the Supervelocity and Selection theory of Light. The rainbow is an analogy for the manifest light bubble. However, one needs to be able to think in velocity space first before they can see the manifest light bubble. The rainbow is a circular, two-dimensional image that can be mistakenly modeled as a real, three-dimensional object, in three-dimensional space. Similarly, the manifest light bubble is a spherical, three-dimensional image, in three-dimensional velocity space, that is mistakenly regarded as a real, three-dimensional object, in three-dimensional velocity space. Because it is not conventional in physics to apply velocity space in this context or to explicitly model light in velocity space, this misconception about light has not been conspicuous enough to either detect or correct.

Instead of making use of three-dimensional velocity space, light is conventionally modeled as a four-dimensional light cone, in a four-dimensional spacetime manifold. Modeling the behavior of light in spacetime is needlessly complicated, in comparison to modeling the behavior of light in velocity space. The main benefit that comes from using velocity space to study light, is that it exposes the many parallels between the observer-centred behavior of the manifest light bubble, and the observer-centred behavior of the rainbow.

Light Speed Cannot Be Relative, Fixed, and Objective

In my experience, I have found that the overwhelming majority of laypeople, and even most professional physicists, do not notice or acknowledge any contradiction in the idea that the speed of light is simultaneously relative to all observers, fixed, and objective. I have also been able to identify two primary reasons for this temporary blindness. One reason is that the relevant language is full of ambiguity and so the necessary linguistic tools have not been made available. The second reason is that the visual and mathematical tools which could assist us with highlighting the contradiction have not been available, either.

I will attempt to make the case that no finite speed (including 299,792 km/s) can have all three of these properties at the same time, for the benefit of those who do not immediately see any contradiction. Fortunately, once the contradiction is seen for the first time, it will become progressively more salient, from that point onwards. I find that the rainbow analogy is also effective at making this latent contradiction more apparent. The rainbow analogy has to do with relative distances and positions, while the speed of light has to do with relative speeds and velocities, but the same overall reasoning applies equally, to both. It is relatively easy to understand how a rainbow can appear to occupy a fixed location or a fixed distance, relative to an observer, without actually doing so. It is much more difficult to understand how light can appear to occupy a fixed velocity or a fixed speed, in relation to an observer, without actually doing so. My hope is that the rainbow analogy, in combination with the visual framework of velocity space, will help to bridge this crucial gap in understanding.

If the speed of light is both fixed and relative to all observers, then the speed of light must also be subjective. Similarly, the apparently fixed and relative position of a real rainbow, also implies that it has no objective distance, or location. A real rainbow is not really in the position that it appears to be in, nor is it really at the distance that it appears to be at. The illusory nature of a real rainbow is the only way that it can maintain a fixed, apparent distance from the observer, no matter where the observer happens to be. Similarly, in velocity space, the velocity of light manifests as a light bubble, with the observer always at the centre of the bubble.

If the speed of light were objective, then the speed of light could either be fixed *or* relative to the observer, but not both, at the same time. While a real rainbow has no objective location, a photograph of a rainbow does have an objective location. A printed photograph of a rainbow can have an objective location that is also fixed, if it has been framed, and placed in an art gallery. Because the location of a rainbow photograph in an art gallery is fixed, relative to the ground, it cannot also be fixed, relative to multiple observers who are in different locations, and who are also free to move around as they please. A printed photograph of a rainbow that is hung in a museum is in an objective location and fixed, but it is not fixed relative to the observer.

Similarly, if the velocity of light were objective, then a beam of light would occupy a single point in velocity space, and this point would not be free to move around, depending on where the observer is in velocity space. This point in velocity space also would not be free to move around, depending on which observer is observing it. Different observers would therefore measure different speeds, and as each observer changes their velocity, the relative velocity of light would change, accordingly.

Now, let's consider the other possibility where a distance is objective, variable, and fixed, relative to all observers. Let's say that this photograph of a rainbow is no longer hanging up in an art gallery, but it has been copied, and distributed as a handheld photograph, to everyone. Each observer is now holding a photograph of a rainbow at arm's length and walking around with it. Now, this rainbow has real, objective locations, and it also has a fixed distance, relative to all observers. But the location of this rainbow is now variable, and it can also have multiple locations, at the same time. In velocity space, this pattern would also look like a collection of multiple dots, each moving around independently, and all widely dispersed across velocity space.

To reiterate, if the speed of light is both relative, and fixed, then it must also be subjective, but if the speed of light is objective, then it must either be variable, or absolute. All of the scientific evidence that we have leads us to conclude that the speed of light is subjectively relative and fixed but objectively dispersed. The way that light actually behaves is analogous to a real rainbow. Light does not behave like the singular photograph that is on display in the art gallery, nor does it behave like the handheld photographs, which are distributed to each observer.

To return to the rainbow metaphor, how would you respond to the claim that the position of a rainbow is a finite and measurable distance away from the observer? What about the claim that the position of a rainbow is always the same fixed distance away from all observers? Our reactions to analogous claims about the fixed and finite nature of the speed of light should reflect what we know about the nature of the rainbow. When similar claims are made about the speed of causality, the speed of gravitational waves, the speed of information, or about the principle of locality, then the only appropriate response is incredulity.

The Bidirectional and Unidirectional Speeds Must Be Different

Just for the sake of argument, let's pretend that we are hopeful and determined to find some way to map the subjective velocity of light onto the objective velocity of light. In order to accomplish this, we would require an explanation that can account for two divergent viewpoints, with a transformation that has two important properties. Firstly, this transformation must be a vector with a directional component. In other words, a scalar transformation cannot work. Secondly, this transformation would need to rely on either addition or subtraction. Our transformation cannot rely on multiplication or division. The two components of a solution that were explicitly proposed by Einstein, in order to account for the otherwise incompatible observations made by observers in different locations in velocity space, are time dilation, and length contraction. But these two operations cannot assist us because what we really need is to add or subtract a vector. Multiplying or dividing by a scalar quantity (as with time dilation) or multiplying or dividing by a vector (as with length contraction) cannot possibly allow us to reconcile these divergent accounts.

Because the rate of the passage of time is a scalar quantity and not a vector, there is nothing that we can do to the rate of the passage of time of an observer to arrive at a coherent solution to this problem. Because slowing down or speeding up time involves multiplication and division, as opposed to addition and subtraction, then dilating or contracting time also cannot be used. Length contraction and length dilation is also not going to work because these are translations that are based on multiplication and division.

If time is to be dilated for a particular observer, then all of their time-based and velocity-based observations must be dilated by the same factor. For example, in order to bring the two observers' observations of the previous bird analogy, into alignment, we could dilate the observer in the car's time by a factor of five, such that the twenty km/h that they would have seen, now becomes one hundred km/h, which was seen by the observer on the ground. But this would not resolve the overall discrepancy because we would also need to multiply the eighty km/h that the observer in the car appears to be moving at, relative to the ground, by a factor of five, in order to get 400 km/h. There is no possible solution that can be achieved with a combination of time dilation and length contraction.

We simply have inconsistent and totally irreconcilable accounts for how fast the bird is moving, relative to various observers, and objects.

This difficulty arises because there are multiple incompatible goals which must be satisfied, at the same time. The reason that multiple discrepancies arise is partly due to the fact that an observer can see multiple objects move relative to them, and also partly due to the fact that an observer can see an object move relative to other objects as well.

There are three reasons which, when taken together, make it obvious that the one-way speed of light is different from the two-way speed of light. The first reason is the fact that the idea that light moves at different relative speeds in different directions, is incorporated into the hidden logic behind the Lorentz transformation equations. Merely using the Lorentz formula to calculate time dilation is itself a tacit acceptance of the idea that light travels at different relative speeds, in different directions. It is not possible to declare that light moves in both directions at the same speed *and* to apply the Lorentz transformations, without contradicting yourself.

The second reason is the fact that Einstein's light clock thought experiment only works when we average across both directions. If the relative speed of light were the same in both directions, then averaging the two directions together would not be necessary.

The third reason is Einstein's own acknowledgment that the assumption that light travels at the same speed in both directions is merely a convenient convention and that this assumption cannot be checked or verified.

The reason that time dilation and length contraction cannot resolve the apparent discrepancies from different inertial frames, is because time dilation and length contraction were never designed for or intended to resolve these discrepancies. Time dilation and length contraction were designed to account for the averaged bidirectional dynamics of light, only.

Einstein on Simultaneity and the Bidirectional Speed of Light

Einstein asserted in *On the Electrodynamics of Moving Bodies (1905)* that the only way to synchronize clocks that are far apart from one another, would be to communicate between the clocks with light signals, and to assume that the speed of light is the same in both directions.

According to Einstein:

> If at the point A of space there is a clock, an observer at A can determine the same values of events in the immediate proximity of A by finding the positions of the hands which are simultaneous with these events. If there is at the point B of space another clock in all respects resembling the one at A, it is possible for an observer at B to determine the time values of events in the immediate neighbourhood of B. But it is not possible without further assumption to compare, in respect of time, an event at A with an event at B. We have so far only an "A time" and a "B time." We have not defined a common "time" for A and B, for the latter cannot be defined at all unless we establish *by definition* that the "time" required by light to travel from A to B equals the "time" it requires to travel from B to A.

Establishing as "true by definition," that light always travels at the same speed, in both directions, from within any given inertial frame, is necessary in order to synchronize distant clocks. The idea that light travels at the same speed in both directions was also a necessary assumption in Einstein's light clock thought experiment and in the development of special relativity. But as I have demonstrated with the longitudinal light clock thought experiment, it is actually necessary to provisionally assume that light moves at different relative speeds, in different directions, in order to perform the basic calculations that determine the degree of time dilation. This is a deep contradiction that requires resolution, and ignoring the one-way speed of light is not the solution. If the one-way speed of light were not only immeasurable, but meaningless, then how could the bidirectional speed of light be meaningful?

If we first assume that the speed of light is independent of its direction, and we also model light's range of potential velocities in relation to a particular location in velocity space, then it necessarily follows that the speed of light would be dependent on its direction, when viewed from all other locations in velocity space. One can visualize this by imagining a manifest light bubble in velocity space that is centred around a particular observer. For any object that is *not* at the centre of this particular light bubble, the distance between that object and the membrane of this light bubble will change depending on the direction that we consider. In other words, some parts of this light bubble would be further away from this off-centre object than others.

The Time-Dilated Observer's Perspective

Let's consider what would be observed from the perspective of an observer who is in a time-dilated frame of reference. We already know that this observer would not perceive their own time as dilated, because they will observe any clock that they are carrying with them to tick at the usual rate. However, the big question is how time in the world outside of the observer's frame of reference will become distorted, according to special, and general relativity, if that observer's time has been relatively dilated.

Let's say that a special drug is given to an observer which will slow down all of their biological processes. After taking this depressant, all of the operations of the observer's mind and body are slowed down to fifty percent of the normal speed. This would be akin to the effect of time dilation. To other people, this observer will appear to be functioning much slower than normal. To this observer, everyone else will appear to be moving and speaking at double the normal pace and with high-pitched, squeaky voices. In essence, the time-dilated observer is slow, in comparison to the environment, and the environment is fast, in relation to the time-dilated observer.

When an observer's time is objectively dilated, this also means that their clock appears to be moving slower from the external reference frame, and that all motion within the observer's frame appears to be moving slower from the external reference frame. Time dilation would also affect the speed at which spheres of light appear to expand, but it cannot alter the shape or the trajectory of these expanding spheres of light. In velocity space, time dilation can only alter the size of the observer's manifest light bubble by increasing the radius of the observer's manifest light bubble.

The Two Opposing Types of Time Dilation

It is not actually possible to answer the question of how time dilation affects the way that an observer perceives the world outside of their time-dilated frame of reference. The reason that there is no single answer is that the effect of time dilation depends entirely on the mechanism behind the time dilation. If the time dilation is caused by gravitation, then the external reference frame will appear to speed up, from the perspective of the time-dilated observer. But if the time-dilation is caused by relative velocity, then the reference frame will appear slower, from the perspective of the time-dilated observer. The reason for this difference between gravitational and inertial time dilation has to do with the presence or absence of what is called "symmetry."

It is very important to clearly distinguish between *gravitational* time dilation and *inertial* time dilation. The reason that it is important to distinguish be between these two types of time dilation is that with inertial time dilation, both observers will disagree on the direction of the time dilation (whose time is relatively dilated and whose time is relatively contracted). However, with gravitational time dilation, both observers will agree on the direction of the time dilation. There is no gravitational twin paradox because gravitational time dilation is considered to be an objective phenomenon. (I will be discussing the twin paradox at length in short order)

Gravitational and inertial time dilation actually cancel each other out, from the perspective of the time-dilated observer. It is only from the perspective of the external reference frame that both types of time dilation have the same effect. With inertial time dilation, the situation is analogous to how two observers both appear smaller to each other as the distance between them grows, even though neither of them is actually smaller. Meanwhile, with gravitational time dilation, the effect is just as real as the objective fact that Earth is larger than the Moon.

The Manifestation of Time Dilation in Velocity Space

To describe how time dilation would manifest in velocity space, we would first need to know whether it is gravitational, or inertial. Gravitational time dilation would manifest as a decrease in the distances between the grid lines in velocity space, from the perspective of the time-dilated observer. This would make everything in the external frame of reference, including light, appear to be moving faster. Inertial time dilation would manifest as an increase in the spacing of the grid lines in velocity space. This would make everything in the external frame of reference, including light, appear to move slower. With gravitational time dilation, all distances in velocity space expand, while the grid itself shrinks, in a relative way. With inertial time dilation, all distances in velocity space shrink, while the grid itself expands, in a relative way.

The Three Distinct Levels of Inertial Time Dilation

The primary stage of inertial time dilation is implicitly used to construct the Lorentz transformation equation, and it changes depending on the direction of the light. The secondary stage of inertial time dilation is an averaged residual rate of change, which we get as an output from the Lorentz transformation equation. The tertiary stage of inertial time dilation is a cumulative age difference, which we get from multiplying the output from the Lorentz transformation, with the total time elapsed, and this might hypothetically manifest as the degree to which two clocks are out of sync. I will refer to this tertiary stage of time dilation as "temporal displacement."

One could imagine that temporal displacement could be erased by bringing the clocks, or the observers, back into the same location, and the same inertial frame of reference. One could also imagine that the temporal displacement might persist long enough to be observed by both observers. The longer the two clocks spend in different locations in velocity space, the more temporal displacement will accumulate, and the more out of sync they should be. The infamous twin paradox is based on the idea of attempting to either systematically observe this temporal displacement, or to prove that it cannot be observed.

The Twin Paradox and Symmetry

The Preliminary Twin Paradox

Let's suppose that observer A and observer B are identical twins who are separated by a distance of 100,000 km/s, in velocity space. Perhaps they were both residents of the same intergalactic space station, who then decided that they wanted to take a high speed trip to different galaxies, at the same time. According to special relativity, observer A's time would appear to be dilated by 6.075 percent from the perspective of observer B, and observer B's time would also appear to be dilated by 6.075 percent from the perspective of observer A. This means that both twins might think that the other twin is aging 6.075 percent more slowly than themselves. Obviously, both twins cannot be correct. But interestingly, while the twins do not agree on which of the two is aging more slowly, the twins do agree on the absolute value of the time dilation. Because the twins have perspectives that are equal but opposite, just as with a mirror image, this situation is described as "symmetrical."

In this preliminary twin paradox thought experiment, both twins agree on what the relative speed difference between them is, and they are both availing themselves of the same Lorentz transformation equation for inertial time dilation, in order to calculate how rapidly the other twin must appear to be aging. Therefore, it makes sense that both twins arrive at exactly the same number. There are two analogies that will come in handy in order to represent this situation. I will refer to the first analogy as the "visual angle analogy," and I will refer to the second analogy as the "poker game analogy." The visual angle analogy represents the interpretation of the twin paradox under SSL theory, and the poker game analogy represents the interpretation of the twin paradox under special relativity.

The Visual Angle Analogy

In the visual angle analogy, the two observers are separated by a fixed distance in ordinary position space. Because each observer is some distance away from the other observer, each observer will also appear to have shrunk, from the perspective of the other observer. As we know, more distant objects always appear to be smaller, all else being equal. The reason that objects look smaller as they recede into the background, is that the visual angle of an object shrinks as that object gets further away from the observer. Both observers will agree on the absolute quantity of the visual angle shrinking because the distance between them is an objective fact. But because both observers understand how this visual angle-shrinking effect works, neither of the two observers will think that the other observer has actually shrunk. Once the two observers are finally reunited, then they will be able to confirm that no shrinking actually occurred.

In the visual angle analogy, you have a reduction in the apparent size of the other observer. This apparent reduction in size is symmetrical for both observers but also reversible and illusory. If both observers were to interpret the apparent shrinking as real, then both observers would be wrong.

In the preliminary twin paradox thought experiment, the symmetrical difference in time dilation also comes with additional logical consequences. By considering both the relative rate of time dilation and the total amount of time elapsed, both twins can also calculate the total cumulative age difference between them. But while both twins will agree on the absolute value of this temporal displacement, both twins cannot agree on its direction. If the total amount of time that elapses is one year, then after one year passes both twins will arrive at the conclusion that the other twin must be twenty-three days younger than themselves. As before, both twins will agree on the absolute quantity of temporal displacement between them, which must be twenty-three days, but both twins will disagree on the direction of this temporal displacement.

The Poker Game Analogy

In a poker game, both players will agree on the amount of money that is at stake, but they will also disagree about who is to receive that money, until the conclusion of the game. The amount of money that is at stake in a poker game progressively increases over time, much like the accumulated age difference between the twins. While the stakes can always be raised in a poker game, they can never be lowered. Similarly, the temporal displacement between the twins can always increase, but there is no possible way to reduce it. The poker game analogy gives us a good way to think about how temporal displacement is modelled in special relativity.

The reason that temporal displacement can never decrease is that there is no such thing as a negative distance in velocity space. A velocity is simply an absolute speed combined with a direction, and the speed component cannot be negative. According to special relativity, one cannot possibly see time contraction due to inertial motion, hence, the lowest possible time dilation occurs when the speed is zero, and there is no way for the speed to be any lower than zero. As differences in age accumulate over time, they become locked in, and they form the new baseline. Temporal displacement is not reversible.

The big question is, does time work more like a poker game, as special relativity says it does? Or does time work more like the visual angle shrinking, as SSL theory says it does? If inertial time dilation is real, then time must work just like a poker game. If inertial time dilation is illusory, then time must work just like visual angle shrinking. Einstein famously quipped, "God does not play dice," in response to the randomized resolution of quantum mechanical symmetry. I would add to this, that God does not play poker, either.

The point of this preliminary twin paradox thought experiment is to convey the idea that the symmetrical effect of inertial time dilation is subjective as opposed to objective. We have no good reason to suspect that either of the two twins is perceiving this situation accurately. There is no need to bring the two twins back together in order to find out which one of them is wrong because it is pretty clear from the outset that both twins must be equally wrong. Because SSL theory says that both twins are wrong about the inertial time dilation, then they must also both be wrong about the temporal displacement of twenty-three days. However, because special relativity says that at least one of these twins must be correct, then the only way to sort this symmetrical disagreement out, would be to bring the two twins back together—without inadvertently doing anything to break the symmetry that has been created. In other words, both players will need to show their cards, so that the winner can be unambiguously identified.

The Standard Twin Paradox

In the standard twin paradox, one twin remains on Earth and the other boards a spaceship, which travels to a distant star at close to the speed of light, and then returns to Earth. The twins eventually meet up again, in order to find out which one of them has aged less. The twins will now be able to directly compare clocks or other visible signs of aging. In order for one of the twins to be right, then at least one of them will have to be wrong. If both twins had been correctly perceiving time dilation, then this would open up a number of questions as to how each twin would be able to maintain their mutually incompatible versions of reality.

The emphasis that is placed on symmetry and on breaking the symmetry in this discourse, underscores the importance that is also placed on subjectivity, and epistemology, as a guide to ontology. It is often claimed that in order to break the symmetry, one of the twins must have been objectively moving, while the other twin was objectively stationary. I have previously identified this misconception as Hydra Head #7 because the arbitrary location where the zero is placed in velocity space should not have any effect on the objective reality of the situation. Neither of the two twins could possibly be "correct" about which one of them was stationary, and this is why I say that bringing the two twins back together is a redundant exercise.

In the poker game analogy, you have a cumulative, real, and irreversible increase in the absolute value of some variable, but the direction of this value is not determined until the symmetry is finally broken. One and only one observer will be correct. In the twin paradox, the winner is the twin who obtains the zero in velocity space, because they can claim that they were not moving, and that it was the other twin that was really moving. The way that this symmetry is broken is that it is claimed that the space travelling twin experienced a force of acceleration, as the spaceship turned around to come back. It is alleged that this force of acceleration is what informed the twin that they were "really" moving.

A better way to understand how the experiences of the two twins are asymmetrical, is by looking at the twin paradox through the lens of velocity space. As seen in velocity space, the earthbound twin remains at the same location in velocity space (roughly speaking), regardless of where we place the zero in velocity space. Meanwhile, the space travelling twin will trace out a pattern in velocity space that looks something like a skinny **S**, **Z**, or figure **8**, regardless of where we place the zero in velocity space. The fact that the space travelling twin has covered much more territory in velocity space is something that both twins can agree on, even if they can't agree on where the arbitrary zero should go. In fact, the question about where the zero belongs is irrelevant because both twins will still have an equal claim to having been "stationary," at any given moment in time.

Even though the twins can both agree on which of the two has remained in the same location in velocity space, and on which of them has covered a large territory in velocity space, it is not clear how this asymmetry can or should be used to select a winner between the two of them. If this difference between them does create a temporal displacement, then this temporal displacement cannot be a result of any difference between them in their inertial motion. Any difference that we find would have to have been caused by either gravity or acceleration. In other words, the acceleration would have to do much more than just pick the winner, which would be very strange. The distance covered in velocity space would also have to determine the absolute amount of temporal displacement.

If we adhere to a consistent understanding of how special relativity says that a temporal displacement should be produced, then the formula to calculate a temporal displacement is very simple. You merely take *gamma,* which is a function of the absolute distance in velocity space between the two observers, and you multiply *gamma* by the amount of time that the two observers spend at a given distance in velocity space. If time is on the **x** axis and *gamma* is on the **y** axis, then the total temporal displacement should be proportional to the area under the curve. However, in practice, this is never how temporal displacement is actually calculated. Instead, the way that physicists have implicitly chosen to calculate temporal displacement for the purposes of testing special relativity, is to calculate the total length of the path taken in velocity space. This is a point that I will return to and expand on in chapter six. As I have mentioned previously, the length of a path taken in velocity space is a measure of the cumulative acceleration or the change in velocity over time. What this tells me is that physicists do not actually use special relativity in order to generate the predictions that are intended to validate the theory.

But just for the benefit of those who think that the primary role of acceleration is to pick the winner or to designate a particular location in velocity space as "stationary," I have devised a more advanced twin paradox thought experiment which will ensure that the force of acceleration is also applied in a symmetrical manner.

The Twin Paradox 2.0

Let's imagine that we have constructed twin elaborate space stations of equal mass. We will send these space stations into orbit around the Sun rather than Earth. The radius of the orbit of both space stations will be 28.26 million km (roughly half the distance of Mercury's orbit) and it will take both of them exactly thirty days to complete one cycle around the Sun. The orbits of both space stations will occupy a plane that is perpendicular to Earth's orbit, but the

But Where is the Rainbow?   149

two space stations will orbit the Sun in the opposite direction, with respect to one another (clockwise vs. counterclockwise). These twin space stations will not crash into or significantly disrupt the orbits of one another because their orbits will also be slightly tilted, relative to the other, by 0.1 degrees.

We can fill up these twin space stations with people and timekeeping devices that work according to different methodologies and mechanics. There will also be an opportunity for easy communication over a relatively short distance between the two satellites, as they make a near pass, once every fifteen days.

Relative to one another, these space stations will be moving at an incredible speed of 137 km/s, as the distance between the two space stations reaches a minimum. This high relative speed would make it possible to observe time dilation as it accrues over long periods of time. It is interesting to note that at the moment when these two space stations are closest together in position space, they are also furthest apart in velocity space, and vice versa.

What we have now is an independent variable which we can manipulate, and that independent variable is the amount of time that the two space stations spend while separated in velocity space. We can now measure and compare any changes between the synchronization of clocks after fifteen days, thirty days, forty-five days, sixty days, seventy-five days, ninety days, and so on.

This thought experiment gets around the problems of the original twin paradox. In the Twin Paradox 2.0, both observers are tracing out identical circles in velocity space, and both twins experience exactly the same amount of acceleration.

This thought experiment could actually be conducted, and if it were to be conducted, then my prediction is that no inertial time dilation will be observed because there is no such thing as inertial time dilation. However, even if there is such a thing as inertial time dilation, then the symmetry would eventually need to break in an arbitrary and unpredictable manner, and not in the way that special relativity says it should.

The Nature of the Illusion of Inertial Time Dilation

According to SSL theory, inertial time dilation is, at most, an illusion, and so the visual angle analogy is more appropriate than the poker game analogy. I say, "at most," because in order for inertial time dilation to be illusory, it would also need to be observed. Because inertial time dilation has only been observed in thought experiments, it would be misleading to call it illusory. In order for a phenomenon to be illusory it must, at a minimum, also be apparent.

Not only do I say that inertial time dilation is not real, but I also say that it is not illusory because it has not been directly observed. The twins could never observe time dilation directly. Instead, what they observe is their relative speed and they both use the same Lorentz equation to calculate what the time dilation would hypothetically be. Hence, the equation functions as a substitute for an observation. If the equation is false, then so is the corresponding "observation." If the equation is not used, then the illusion never appears.

The entire "illusion" was created merely by inappropriately applying an equation. This equation is itself based on a false assumption about light, namely, the notion that all observers see the same light. While SSL theory says that there is no inertial time dilation, it has nothing to say about the existence of gravitational time dilation, time dilation caused by acceleration, or general relativity.

Approaching the Speed of Light From Different Angles

Superwave Dynamics

The Pentality of Light

I mentioned in the introduction that the SSL theory does not propose any wave-particle duality. The idea of wave-particle duality is somewhat misleading because it is also oversimplified. I prefer the term *pentality* over duality because it appears as though light actually manifests in the following five distinct forms:

1. Particle
2. Pair of orthogonal transverse waves
3. Longitudinal wave
4. Helix
5. Superwave

At the most superficial level, light resembles a particle because it sometimes appears to have properties such as position and momentum. At a deeper level of analysis, light resembles a pair of transverse waves and also appears to have perpendicular polarity, phase, and frequency. Light also behaves like a pressure wave, as it exhibits spherical diffraction, amplitude without spatial displacement, and ambiguous polarity. Light also resembles a helix, from within the established mathematical framework of quantum mechanics. Finally, when considering the speed of light, light also resembles a superwave, meaning that its velocity is in superposition by default: Light spreads out across velocity space. Because it has both positive and negative speed simultaneously, it is also able to travel in opposite directions at the same time.

SSL theory provides us with an important insight about light which is worth repeating here. This insight is that light appears to travel in multiple directions at once, *because* it travels at multiple velocities at once. The fact that light is a superwave also explains how light can display some of the characteristics of longitudinal waves while also displaying some of the characteristics of transverse waves; it explains why photons can possess multiple polarization vectors at the same time, and it explains why the amplitudes of light waves have no measurable distance. Light is not a wave at all, it is a superwave.

The Helix

A helix is a simple coil, spring, or corkscrew pattern. Helical motion combines circular motion with constant motion that is perpendicular to the circular motion. When viewed subjectively from specific angles, a helix may resemble either a circle or a wave.

Light is modeled as a helix in quantum mechanics because it is convenient to do so. Imagine that you wanted to create a cartoon of a moving wave as efficiently as possible and project that onto a screen. You could try drawing hundreds of waves by hand, but this would be both difficult, and time consuming. Fortunately, there is an easy way to accomplish this. If you take a wire and bend it into the shape of a helix, then all you need to do is rotate the helix, while holding it at a right angle to a light source and projecting a shadow of the helix onto a screen. This shadow would resemble a wave. You can then make this wave image move smoothly and continuously at any speed just by rotating the helix at different speeds and in different directions.

Mathematicians and physicists who create mathematical models of waves have also found it convenient to use an analogous method.

Using a three-dimensional helix in order to create a two-dimensional projection of a wave, allows mathematicians to arrive at relatively simple, and elegant equations to describe the full range of dynamics of a wave, over time. This is why it has become customary to model wave dynamics in quantum mechanics with a helix.

This does not necessarily mean that the helix represents a deeper reality that we can only observe indirectly. On the other hand, this might be an indication that there really is a latent helix which only manifests as a wave when we observe it. According to the mathematics of quantum mechanics, the helix is only visible when the imaginary (*i*) dimension is included in the representation. When only real numbers are considered, the helix disappears, and we are left with only a wave shape.

Transverse Waves vs. Longitudinal Waves

Transverse waves are the sort of waves that we see on the surface of the water. The direction of the vibration, or the polarization of a transverse wave is always perpendicular to the direction of motion. By contrast, the polarization of a longitudinal wave, such as a sound wave, is always in alignment with the direction of motion.

Because an extra dimension is required to support the polarity of transverse waves, transverse waves can only travel in a maximum of two dimensions. This is why ripples in a pond form two-dimensional expanding circles. These waves can move in all directions horizontally, but the vibrations occur strictly along the vertical axis. By contrast, because sound waves don't require an additional dimension for their polarization, they are free to travel in all three dimensions.

Light waves actually use all three dimensions for their motion in a way that is elegant, beautiful, and enigmatic. Light waves use one dimension to support the electric component of their polarization, a second dimension to support the magnetic component of their polarization, and a third dimension for their motion. One would therefore expect that light waves can only move in a single direction at any given time, just like a particle.

However, the reality is that light waves generally move in all directions in three-dimensional space, at the same time. The manner in which light waves move is identical to how longitudinal waves move, even though their polarity is akin to a dual transverse wave. This set of facts is fascinating but also very difficult to make sense out of.

The received wisdom is that the amplitude of light waves has no spatial meaning. In contrast to the wavelength of light, it is not possible to specify the degree of displacement of a light wave in units of distance. Instead, the amplitude of light is always measured in units of energy. This, again, does not appear to make any sense. If a wave has a defined polarization, then it must also have a vibration, which has a magnitude that can be measured in terms of distance. A thing cannot move in a specific direction without also moving a specific distance. Both transverse waves and sound waves can be measured in terms of the real, physical distance of their vibrations, and so why not light waves? Light waves also have real, physical wavelengths, so why wouldn't they also have real, physical wave amplitudes? At this moment, these basic questions have no official answers.

The Necessity of Superposition

Electromagnetic waves expand in all directions simultaneously in a coherent, spherical structure, just as sound waves do. But unlike sound waves, light waves also exhibit polarity in their electric and magnetic components, which are perpendicular to the direction of motion, and also perpendicular to each other. Unlike with surface waves, the direction of the polarity within the perpendicular plane is arbitrary, and it normally exists in superposition, which means that it is not determined, or specified in advance.

So long as the polarity of the waves remains in superposition, then there is no visible problem or contradiction, and the overall structure does not need to make sense or be geometrically coherent. But if we imagine that at each point along the spherical surface structure of the expanding wave, there actually does exist a defined polarity for both the electric, and the magnetic components, then how can we model this in a way that makes sense?

You can try to imagine drawing hundreds of tiny arrows, at hundreds of points, along the surface of a sphere, to indicate the polarity of the wave at each point on the surface. How can these arrows be drawn such that they are all the same magnitude, and there are no special locations, such as poles, anywhere on the surface of the sphere? How can the local property of polarity be reconciled with the global geometry of the sphere? There does not appear to be any way to draw these arrows such that they are aligned with one another without creating distortions, unevenness, and special locations on the sphere.

This may explain why superposition is necessary as a way to mask the contradictory and incoherent nature of the polarity of electromagnetic waves. There does not appear to be any global structure to the polarity of spherically shaped light waves. This paradox was pointed out by Henri Poincaré with his hairy ball theorem, over one hundred years ago.

Frame-Dependent Direction of Motion

The Angle and the Light Clock

With respect to the light clock thought experiment, only the length of the apparent path taken by the light is normally considered to be relevant. However, the angle at which the light is reflected between the mirrors also changes, along with the relative motion of the light clock. This angle variable is ignored because it is not relevant to calculating time dilation or length contraction, but it is no less interesting. What looks like light bouncing back and forth at a ninety degree angle, to the stationary observer, might look like light bouncing back and forth at a sixty degree angle, to the moving observer.

The implication of this is that it is not only the speed component of the velocity of light that is affected by relative motion. The direction component of the velocity of light is also affected by relative motion. In theory, one should be able to change the direction that light moves in, simply by accelerating in one direction or another.

The Photon as an Arrow

Imagine photons as tiny arrows! An arrow should always point in the direction of its motion. But because motion is relative to the frame of reference, then as the frame of reference shifts, the direction of the motion of the arrow also shifts. Therefore, the frame-dependent velocity vector of the arrow may point in a different direction as compared to the frame-independent direction that the arrow is pointing in.

Light has analogous arrowlike properties. The polarization of the electric and magnetic oscillations of light always point at right angles to the direction of the motion of the light wave. But if the direction of the velocity of light is determined by the observer's location in velocity space, then does the location of the observer in velocity space also determine the direction that the electrical and magnetic arrows point in?

This is a question about how the polarization of light is affected by the relative motion of the observer. With the light clock thought experiment, one can imagine that the arrow of the light is pointing at an orthogonal angle to the mirrors, while the apparent motion of the light is pointed at a different angle. The implications of the hairy ball theorem appear to lead to the conclusion that photons should not have any defined polarization until they are observed. Hence, their polarization can be defined in a post hoc manner, after their velocity vector has already been defined. The fact that the polarization of light is in a superposition of all possible values by default, is precisely what quantum mechanical experiments have confirmed.

Applying the Analogy of the Visual Angle

Visual Angle, the Moon, and Symmetry

The introduction of the concept of *symmetry* to theoretical physics was also the tethering of ontology to epistemology and the introduction of the strange idea that epistemology guides and shapes the laws of the universe. Every known symmetry is also regarded as a limit on knowledge. According to Richard Feynman and Herman Weyl, "A thing is symmetrical if one can subject it to a certain operation and it appears exactly the same after the operation." (*Six Not-So-Easy Pieces*, 1963) According to Emmy Noether's influential theorem, "Every differentiable symmetry of the action of a physical system with conservative forces has a corresponding conservation law." (*Six Not-So-Easy Pieces*, 1963) In other words, if you can't tell the difference between two possible states then there must be a symmetry there, and if there is a symmetry there, then there must also be some underlying conservation law, which makes it so. But there is something unsettling about elevating a mere appearance to the status of an immutable law of the universe. It there really a cosmetic cosmic dress code?

The symmetry that forms the basis of Galilean relativity is the notion that it is not possible to determine the absolute velocity of anything. The symmetry that forms the basis of special relativity is the notion that it is not possible to use the relative speed of light to determine the absolute velocity of anything. In other words, all locations in velocity space are on an equal epistemological footing.

The symmetry that forms the basis of general relativity is the notion that it's impossible to tell the difference between the influence of an accelerating frame and the force of gravity. In all cases of symmetry, it is strongly implied that because it is impossible to tell the difference between two things, then this must also mean that there really is no difference between them. Epistemological limitations are assumed to come with ontological ramifications under this paradigm.

There is also a unique symmetry that is associated with the Moon, and this has long presented humanity with a profound mystery. Despite the Moon's prominent presence, there was never any way for our ancient ancestors to determine how large or how far away it was. One could try a blind guess, but without the assistance of advanced science, a blind guess could have been wildly inaccurate. We are not naturally equipped with the advanced tools that are necessary in order to gain a working perspective on astronomical proportions, and the Moon provides a particularly striking example of this ineptitude.

We do have other methods of depth perception available to us, however. With our innate depth perception through the use of binocular stereopsis, it is possible to determine that the Moon can be no less than twenty meters away. Through adjusting the focus on the lens of the naked eye to reduce the apparent blurriness of the image of the Moon, it is possible to determine that the Moon can be no less than thirty meters away. However, these techniques do not allow us to determine any maximum distance, or to narrow our estimate down any further. We could use the occlusion of clouds to refine our estimate of the distance to the Moon, but then the distance to the clouds would also need to be estimated. From the naïve perspective of a hunter-gatherer observer, the Moon might as well be infinitely far away, or perhaps only 10 kilometers away, or anywhere in between.

The visual angle of the Moon, as viewed from Earth, remains fairly stable at roughly one half of a degree. If you imagine your total field of view as being a spherical projection screen that surrounds you, then a visual angle is just a practical way of measuring, and describing how much of that imaginary screen is taken up by a particular object, due to a combination of its size, and its proximity to you. To put it plainly, the visual angle is a measure of how big an object looks from where you are standing. In the case of the Moon, the visual angle had created a symmetry, with respect to both size, and proximity because it was impossible to tell how much each of these two factors had contributed to the visual angle of the Moon.

Size and proximity both resemble one another because latent changes in both variables produce a manifest change in the visual angle. However, we generally don't have much trouble with determining the size and distance of objects because we have many other ways to estimate these variables. Firstly, we are capable of gathering multiple perspectives on a single object from different locations and comparing these observations. Secondly, we are able to use multiple methods in addition to the visual angle in order to arrive at independent estimates for both size and distance.

Once you know the size of an object, then you can easily determine the distance by interpreting the visual angle information in a new light. Similarly, once you know the distance of an object, then you can also determine its size by interpreting the visual angle information, in light of the known distance. There are also several important heuristics that we rely on, in order to make sense out of the visual angle information, such as the idea that similar objects tend to have a similar size, that identical objects tend to have the same size, and that objects rarely, if ever, change their size, without also changing their overall appearance. In other words, we get accurate results when we apply the assumption that size is a relatively fixed property of an object, while distance is relatively variable. It is partly thanks to our instinctive application of powerful heuristics like these, that we are rarely confused about how large or how far away the objects in our immediate environment are.

So, why not employ similar heuristics and methodologies in order to unravel the mysteries of modern physics? A visual angle is a mere appearance. A visual angle is superficial. A visual angle is further towards the subjective and manifest end of the spectrum, as opposed to the objective and latent end of the spectrum. Similarly, the speed of light is also a mere appearance. The speed of light is much further towards the subjective and manifest end of the spectrum, and the deeper and more coherent reality that manifests as the speed of light, has been waiting to be unearthed.

Terrestrial objects change their apparent size, depending on where the observer is standing, but the size of celestial objects like the Sun and Moon, remain invariant. Similarly, ordinary objects change their apparent velocity, depending on the motion of the observer, but the apparent velocity of light remains invariant. I would argue that just as we don't accept the visual angle as a fixed property of any object, we also should not accept the idea that the time it takes for a beam of light to travel a fixed distance is a property of the beam of light.

For relatively small and distant objects, visual angle can also be represented as a family of related equations, where the visual angle is equal to the size of the object, divided by its distance from the observer.

$$\text{Visual Angle} = \text{Size}/\text{Distance} \quad \text{Size} = (\text{Visual Angle})(\text{Distance}) \quad \text{Distance} = \text{Size}/\text{Visual Angle}$$

These three equations show you how to calculate the third variable if you have the other two. From these equations, we know that an increase in the visual angle could reflect either an increase in size or a decrease in distance. An increase in size might either manifest as an increase in the visual angle or as an increase in the distance of the object. We also know that an increase in the distance of an object will either manifest as an increase in the size of the object, or as a decrease in the visual angle.

Because we already have an understanding that the visual angle is not as stable as the distance, which in turn is not as stable as the size, then confusion only manifests when all we know is the visual angle, but we don't know either the size, or the distance. The hierarchy of stability for the visual angle would look something like this:

Size > Distance > Visual Angle

Visual angle is the least stable and reliable of the three, which is why it seems so strange to us that a visual angle could be stable, such as with the Moon. Distance is in the middle because objects and observers tend to move around. Finally, size is the most stable of the three properties. From this hierarchy of stability, we can easily derive the helpful heuristics that I mentioned earlier. We can assume that objects tend to maintain a consistent size, for example.

I think that it would also be useful to develop and adopt a hierarchy of stability for the velocity family of equations, to compliment the one that we have always relied on for the visual angle family of equations. That hierarchy of stability would look something like this:

Time Elapsed > Distance Covered > Velocity

Based on this hierarchy, we should assume that the time elapsed is the most stable and trustworthy value and that velocity is the most variable and unreliable. Regrettably, special relativity implicitly inverts this hierarchy. Within special relativity, the velocity of light is considered to be the most stable, distance has intermediate stability, and time is considered to be the least stable of the three.

While the oversimplified visual angle and velocity equations that we have been using work well for small visual angles and slow speeds, it appears as though they are not adequate for large visual angles, and fast speeds. Let's explore what this means with another analogy.

Visual Angle, Earth, and Infinity

In the two sections that follow, I will be provisionally accepting the overall paradigm of special relativity, including both time dilation, and length contraction. This provisional acceptance is necessary, in order to explore special relativity from different angles, in order to gain familiarity with the theory, and to uncover its less obvious implications. This should not be taken as a concession on my part, which would obviously contradict the Supervelocity and Selection theory of Light that I have proposed.

As the visual angle of an object increases, there is also an interesting distortion that occurs. This distortion occurs because there is a maximum visual angle that can manifest, even though there is no minimum, per se. The maximum visual angle is always 180 degrees. A perfectly spherical object cannot cover more than half of the visual field. This is why a scene typically appears to be half ground and half sky. The visual angle of Earth is at the limit, for all intents and purposes, if you are standing on the ground. This means that with respect to the visual angle of Earth, you would not notice any difference if you were to get closer to Earth, and you also wouldn't notice any difference if Earth were to grow larger. From your standpoint on the ground, Earth might as well be infinitely large, and perfectly flat. The distortion that becomes increasingly obvious as you approach 180 degrees is remarkably similar to the distortion that occurs, according to special relativity, as the relative velocity of an object approaches the speed of light.

180 degrees of visual angle is what a spherical object of infinite size looks like, relative to an observer, at any distance. 180 degrees

of visual angle is also what a spherical object of any size looks like, relative to an ideal observer, at an infinite proximity. Therefore, 180 degrees of visual angle represents either infinite size or infinite proximity, or both infinite size and infinite proximity. Similarly, according to special relativity, **c**, which is a finite value, is what an infinite relative velocity would look like, to an observer.

Earth's visual angle is constant at 180 degrees, but it is *not* really the size of Earth that is special. Rather, it is the visual angle of 180 degrees that is special. By the same token, according to special relativity, it is not the speed of light itself that is special, but it is the speed of 299,792 km/s, that is special. Just as 180 degrees is a universal size limit, 299,792 km/s is a universal speed limit. Just as the limit of 180 degrees is a cosmetic law, so is the speed of light, because these are both based on mere appearance. Just as it sounds nonsensical to claim that no object can be larger than 180 degrees, it should sound equally nonsensical to claim that no object can move, relative to an observer, at faster than 299,792 km/s. Just as 180 degrees is not actually a size, 299,792 km/s, relative to an observer, is not actually a speed. And lastly, just as a visual angle is a combination of both proximity and size, a speed is a combination of a distance in space, and a distance in time.

In essence, whether a planet is the size of the Moon, the size of Jupiter, or the size of 100 Jupiters, its visual angle will always be (roughly) 180 degrees, from the perspective of someone who is standing on its surface. By the same token, a beam of light will always appear to be moving at **c**, relative to an observer, no matter how fast it is actually moving, relative to that observer. According to special relativity, the apparent velocity of an object also becomes useless for determining very large velocities, with precision. For example, there is an enormous difference between moving at ninety-eight percent of **c** and moving at ninety-nine percent of **c**, but it is very difficult to perceive the difference between these two speeds.

To someone who is observing from sea level, Earth looks perfectly flat, and infinitely large. The same is true for an observer who is standing on a hilltop. But we don't therefore conclude that that Earth must be infinitely large and perfectly flat, just because it *appears* to be. Similarly, just because the speed of light appears to be **c**, relative to us, no matter what our velocity is, this doesn't mean that its velocity really *is* **c**. In some sense, it may actually be true that the speed of light is 100 times the value that we attribute to it, but because we don't have the capacity to make such a fine distinction at such speeds, then all speeds that are above a certain threshold will look equivalent.

Visual Angle and the Three Speeds of Light

Imagine that you only understood Newtonian physics and that based on this understanding, you knew precisely how much energy it would take to accelerate a cannonball to 299,792,458 m/s, which is also the speed of light. Somehow, you were able to build a super cannon that is capable of launching a cannonball at this incredible speed, into space. You haven't taken special relativity into account with any of your calculations, up until this point. You just know that the more force you apply to the cannonball, the faster its speed will be, in the manner prescribed by Newtonian mechanics. You therefore assume that the speed of the cannonball will be proportional to the amount of momentum that it carries. The more kinetic energy you add to the cannon ball, the faster it will travel.

But if you apply the equations of special relativity, then you may be a little disappointed with the result. Instead of 299,792,458 m/s, you should only see the cannonball travel at 211,985,280 m/s. In other words, from our perspective on Earth, the cannonball will appear to move much slower than we would expect it to, given how much force we have applied to it. However, the impact and explosion that you observe from Earth will appear to have just as much force as you would expect it to, under Newtonian mechanics. These are not contentious claims. This is what special relativity says must happen, even though it is impossible to conduct controlled experiments with such extreme quantities of energy.

The square root of two and its inverse are very important numbers in special relativity. However, these two numbers are important in a way that is scarcely discussed. There are actually three special speeds that are closely associated with light, and those are infinite m/s, 299,792,458 m/s, and 211,985,280 m/s. 211,985,280 m/s is roughly equal to **c** divided by the square root of two. While these three speeds are each distinct, they are also easily mistaken for one another, due to both the ambiguities of language, and the confusion between appearance and reality.

According to special relativity, infinite speed must appear as 299,792,458 m/s, and 299,792,458 m/s, must appear as 211,985,280 m/s. Therefore, **c** might refer to any one of these three special speeds that are associated with the speed of light, depending on the context. The appearance never lines up with the reality, but yet you can always attempt to map the appearance onto the reality, with an equation. Mapping appearances onto reality is precisely what the Lorentz transformation equations are intended and purported to do in special relativity. I have decided to call the appearance of speed the *manifest speed* and the corresponding reality of speed, the *latent speed*.

If we return to the visual angle analogy, we can imagine what angle would correspond to a speed of 211,985,280 m/s. This visual angle is actually an irrational number (115.03672681894...) but I will round it off to 115 degrees, for the sake of brevity. A comprehensive comparison between the three speeds and the three visual angles is provided in the chart below.

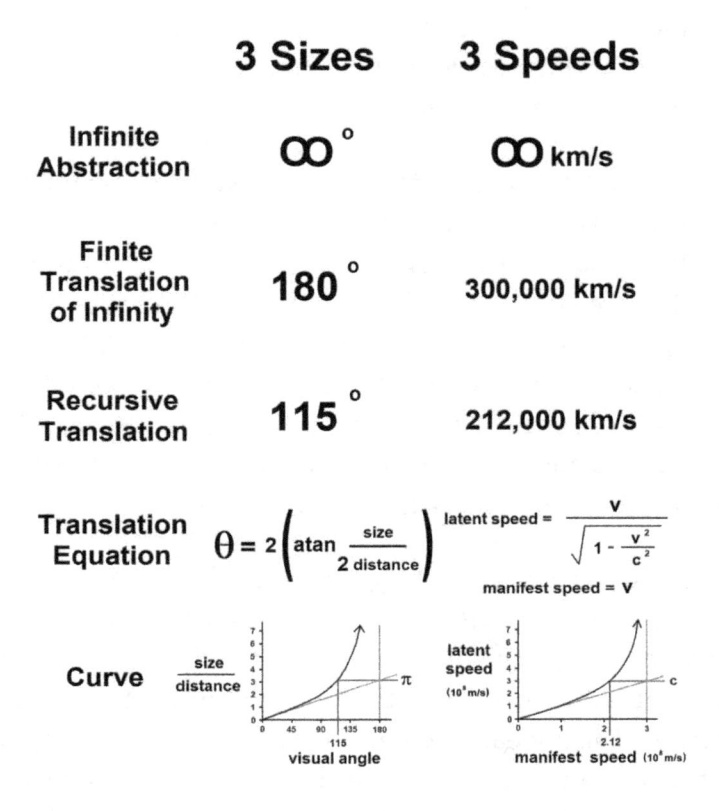

	3 Sizes	3 Speeds
Infinite Abstraction	$\infty\degree$	∞ km/s
Finite Translation of Infinity	$180\degree$	300,000 km/s
Recursive Translation	$115\degree$	212,000 km/s
Translation Equation	$\theta = 2\left(\mathrm{atan}\dfrac{size}{2\ distance}\right)$	$\text{latent speed} = \dfrac{v}{\sqrt{1 - \dfrac{v^2}{c^2}}}$ manifest speed = v
Curve	$\dfrac{size}{distance}$ vs visual angle	latent speed vs manifest speed

115 degrees is what the projected value of 180 degrees looks like once the objective to subjective translation is considered. This means that if you expected the change in visual angle to continue to behave consistently, as it does for distant objects which have a small visual angle, then you would be led to expect that at a distance of 1,068 km, the moon would appear to have a visual angle of 180 degrees. But due to the angular shrinking effect that becomes significant at close proximities and large sizes, the visual angle of the Moon would only be 115 degrees, at that distance. Angular shrinking is analogous to the effect of time dilation or length contraction.

In the cannonball example, special relativity predicts that we would see a massive explosion on the Moon precisely 3.094 seconds after the cannon fired. This considers the time that it takes for the cannonball to reach the moon, from our perspective, plus the time that it takes for the light from the impact on the moon to reach us. No matter how much kinetic energy we put into that cannonball, it will never be possible to see an explosion occur less than 2.5 seconds after the cannon fires—at least not according to special relativity.

It might help if I also frame this cannonball thought experiment mathematically in terms of the Lorentz equation for relativistic mass increase. The question is, for what values of **v** (the relative velocity as expressed as a fraction of the speed of light) and gamma (the amount of relativistic mass increase) would **v** be the inverse of gamma? It turns out that **v** is the inverse of gamma when gamma is equal to the square root of two. This means that the apparent decrease in velocity perfectly mirrors the apparent increase in mass at this special velocity.

Why does gamma need to be the inverse of the velocity? Because one is the only number which is equal to its inverse, and under Newtonian mechanics the value of gamma would always be equal to one, even if **v** were also equal to one (which would mean that **v** is equal to the speed of light). So, the square root of two is the only value for which the product of these two values is exactly equal to one, just as they would be if **v** was equal to **c** and if Newtonian mechanics was correct.

The main take-home point from these comparisons between apparent velocity and visual angle is that just as relying on the visual angle is a relatively shallow, subjective, and unreliable method for measuring distances, so is relying on the apparent relative velocity of an object, according to special relativity. Implicit in special relativity is the idea that observed velocities are never what they appear to be. Infinite velocity makes just as much sense in special relativity, as infinite visual angle does in optics. The limit on velocity that is hypothesised by special relativity is a purely subjective limit and there is no corresponding limit for the objective reality.

What is one percent of infinity? According to special relativity, a cannonball with infinite momentum would move at a finite, relative speed of 299,792 km/s. Because one percent of infinity, is also infinity, then if that momentum were reduced by ninety-nine percent, then that cannonball would still travel at a finite, relative speed of 299,792 km/s. Therefore, this cannot be what is meant by *one percent of the speed of light*. Even though momentum, kinetic energy, and speed are always relative, special relativity treats momentum and kinetic energy as latent properties, while speed is treated as a manifest property. Special relativity regards speed as being just as superficial and unstable as visual angle is. This is because special relativity explains the behaviour of light as resulting from the geometrical structure of space and time, and as a general property of speed.

By contrast, SSL theory maintains that the behaviour of light is radically different from the behaviour of matter, and that the behaviour of light cannot be explained as a consequence of the geometrical structure of space and time. If we adopt SSL theory, then the apparent magic and mischief that we see in the behavior of light would not spill over into more general physical concepts. SSL theory treats light as the exception, rather than as the rule, such that speed, space, and time, can all be regarded as stable, objective, and real.

Special Relativity as Obtuse Optics

To summarize what I see as misleading and confusing about special relativity, I have constructed an allegory about a fictional theory. Special relativity is a theory about how the observed relative speed of light depends on the location of an observer in velocity space. *Obtuse Optics*, on the other hand, is a fictional theory about how the observed relative direction of light depends on the location of an observer in position space.

One can try to imagine having a very naïve conception of optics, looking up at the stars, and being puzzled and amazed by the strange behavior of the starlight. Let's say that you are observing a particularly bright and distant star, such as Deneb. You also imagine that all of the light from Deneb is able to find its way to your pupil because the beam of light that emanates from Deneb appears to be aimed directly at your pupil, just like a laser beam.

How do you know that all of the light from Deneb is directed at your pupil? Well, you don't actually know that it is. This is merely a naïve assumption. But you do know that all of the light that you can see coming from Deneb is also light that just so happened to be directed at your pupil. All of the rest of the starlight, if it exists, is totally invisible to you. Having no direct evidence of its existence, you find it reasonable to conclude that the rest of the light probably doesn't exist.

You also notice that something strange happens when you decide to change your position. As you take a step to the left, the starlight appears to also change its direction, but only slightly, such that it can still find its way to your pupil. How exactly did the starlight know to change its direction by just the right angle? How is the starlight able to track and find both of your pupils so that it can travel in a straight line, directly from the star, and to each of your pupils? How can the same starlight possibly travel at two different angles, at the same time?

In the real world, we do know the answer to these questions, but only because we have intuitively internalized some of the basic principles of optics. We know that there is actually much more starlight than what we can see. We know that what we see when we look at a star is not really the starlight that it emits, but rather the star itself, and only as seen from a particular angle. We also know that even though it may appear as though the starlight is travelling in a straight line from the source to the observer, the starlight is actually travelling in all directions at once. The narrowing down of the direction of the motion of the light only occurs by virtue of the fact that the observer is in a particular location which excludes all of the other possible locations that they could be in. While the starlight that we observe directly does actually travel in a single, specific direction, the reality is that the starlight doesn't possess or choose any particular direction. The observer acts as a selector, by virtue of their specific location. The act of selecting a particular direction for the observed starlight to travel in is accomplished by the observer, not by the star.

If we didn't have these essential insights about the principles of optics then we might be in search of an explanation for how the direction of the starlight is determined, for why this direction changes, as the observer changes their location, and also for why this direction can appear to be different, for different observers, who are in different locations.

Let's say that to meet our demand for an explanation, a brilliant thinker arrived at a bold, new solution which shocked the world. The solution was a theory that has been dubbed "Obtuse Optics" (pun intended). According to Obtuse Optics, the angle of a beam of starlight is always determined mathematically, by a straight line drawn between the source, and the observer. Also, according to Obtuse Optics, the direction of the starlight is always pointed towards the observer, irrespective of where the observer happens to be standing. Furthermore, the angle of the starlight is also always directed towards the observer, regardless of where the star (the source) is located in the sky.

Obtuse Optics actually fits somewhat well with common sense. Everybody knows that a star cannot move in multiple directions, at the same time. The same is true of an observer. Almost all of the objects that people interact with on a daily basis, can only move in one specific direction, at a time. Moving in multiple directions at the same time just wouldn't make any sense. But according to Obtuse Optics, not only does light only move in one particular direction at a given time, but it can also only move in a straight line, from the source to the observer.

In order to accommodate the new postulates of Obtuse Optics, the foundations of geometry needed a few renovations. While it appears that the starlight is leaving from the star at two different angles, this cannot be the case, as that would create a contradiction. We still assume that the starlight can only travel at one precise angle. The starlight is leaving from the star at one angle, but it is somehow simultaneously directed at multiple observers. To allow this to happen, each observer must experience a compensatory angular shift. According to the brilliant thinker, what must happen is that the observer's perspective also rotates slightly, and imperceptibly, as they move to the left or to the right.

In other words, an observer at location A sees the light moving at angle A. Meanwhile, an observer at location B sees the light moving at angle B. There is a difference between these two angles which should not exist, according to Obtuse Optics. In order to make these two angles equal, we simply rotate the orientation of each observer by a precise amount, such that the two divergent angles end up in agreement. This angular rotation operation does, of course, create a number of new problems, and paradoxes to solve. If you want to know how to solve these additional problems and tie up all of these loose ends, then you can study the subject matter intensively for another ten years, and fully train your mind to think within this new optical paradigm. The mathematics will be extensive, detailed, precise, convoluted, and counterintuitive. You will be trained to apply epicycles on top of other epicycles that have been added, in order to patch up the inadequacies of previous versions of the theory.

In case you are wondering, the angular rotation which takes place at the observers end, is intended to be analogous to the twin concepts of time dilation, and length contraction. As with time dilation and length contraction, angular rotation does patch up a few contradictions, but not nearly as many as it creates. There is a much more parsimonious explanation for the divergence in observed angles. The simple answer is that light does travel from A to B but it doesn't really have an angle of motion. Light only appears to have an angle of motion from the perspective of a single observer. The deeper reality is that light travels in all directions, at the same time.

If travelling at extremely fast velocities were just as common as it is to move from location to location, and if the speed of light were as obvious as the direction of light is, then we would have the sufficient intuitive grounding to see the inherent problems with special relativity. Because we don't have life experience with moving at light speed or with measuring the speed of light, then we cannot evaluate the plausibility of special relativity with the help of common sense. If we did have that experience however, then special relativity would seem just as absurd as Obtuse Optics does. The only insight that needs to be gleaned from this thought experiment, is the idea that light doesn't really have a particular speed. When light moves, it actually moves at all speeds, simultaneously. There is something about our inertial frame of reference, or our location in velocity space, that filters out all of the other light which doesn't match the narrow range of speeds that we are able to interact with and observe.

Empirical Evidence and Predictions

Falsifiability and Risky Predictions

I have a adopted the view of scientific progress that was dominant in the twentieth century, after having been established by Karl Popper, and Thomas Bayes. According to this view, the currency of scientific credibility is only earned through prior exposure to the danger of genuine attempts to refute a given theory. Theories that have been protected from the risk of refutation have less scientific credibility, while theories that have been exposed to real challenges, and have survived attempted refutations, will gain enhanced scientific credibility.

According to Karl Popper's view, a scientific theory is characterised by risky predictions, and a theory that cannot be refuted is one which is unfalsifiable, and therefore pseudoscientific. Related to this is the idea that a useful explanation must also be a useful prediction, and that the reward of enhanced credibility must be tied to the risk of refutation. Along similar lines, Thomas Bayes introduced a rigorous mathematical, and statistical framework, where one updates the probability that a theory is true, over time, through a process of conjecture, and refutation. It is no accident that Bayesian probability theory has been applied to many fields, including stock market trading.

For precisely the same reason that a gambler is not permitted to place a bet on the outcome of a sports game, *after* the result has been determined, and publicized, a theorist should not be permitted to determine what results their pet theory predicts, *after* the relevant data has already been collected, and analyzed. To the extent that the advocates of a theory are permitted to cheat, by having access to the evidence in advance, the status of that theory must be demoted from scientific, to pseudoscientific. If what a theory both predicts and forbids is not perfectly clear and precise *in advance* of a consideration of the evidence, then it must not be considered to be perfectly clear and precise, *after* a consideration of the evidence, either. Making use of the data, prior to deciding how to interpret both the theory, and the data, is akin to engaging in insider trading.

Regrettably, it has become trendy in science to talk about the *prediction* of a theory when the more appropriate term would be *post-diction*. A post-diction would be like a prediction, except that it does not manifest until after all of the relevant evidence has come in. A post-diction is only able to account for what we already know because of its impressive hindsight. In recent decades, the word *predicts* has come to be used as a shorthand way of saying *is consistent with*. In other words, the theory is consistent with the observations, even though the observations were never actually predicted in advance. The effect of this misuse of the word *prediction*, is to give unfalsifiable, and unscientific theories, an air of scientific credibility which they do not deserve.

Karl Popper also provides an insightful and eloquent account of how formerly scientific theories can transform into pseudoscience. According to Karl Popper in *Science as Falsification* (1963):

> Some genuinely testable theories, when found to be false, are still upheld by their admirers – for example by introducing ad hoc some auxiliary assumption, or by reinterpreting the theory ad hoc in such a way that it escapes refutation. Such a procedure is always possible, but it rescues the theory from refutation only at the price of destroying, or at least lowering its scientific status.

As we will see, all of the evidence that has thus far been amassed to support inertial time dilation and length contraction comes from a combination of circular motion and circular reasoning. I will also demonstrate that SSL theory is a genuinely scientific theory by the standards of Karl Popper. SSL theory makes clear, testable predictions, and it is falsifiable.

Empirical Evidence for Time Dilation and Length Contraction

Magnetism and Length Contraction

An electromagnet has been frequently cited as an experimental verification of length contraction. Both Edward Mills Purcell and Richard Feynman have popularized the idea that the previously observed phenomenon of magnetism can now be retroactively explained as a consequence of special relativity. Magnetism is often falsely cited as a prediction of special relativity, even though it is merely, at best, a post-diction.

The basic idea is that if you have negatively charged electrons moving relative to positively charged atomic nuclei, from the frame of reference of the electrons, the atomic nuclei are compressed together, in the direction of relative motion. This compression makes it appear as though there are more atomic nuclei than electrons and therefore, there must also be an overall positive charge. Meanwhile, from the frame of reference of the atomic nuclei, it is hypothesized that the electrons must appear to outnumber the atomic nuclei, resulting in an overall negative charge. In other words, you can obtain both a positive and a negative charge from nothing, and this would necessarily violate the principle of the conservation of electric charge.

But let's imagine that this account is true and that because space has shrunk in the direction of motion, a larger number of particles will appear to fit into the same amount of space. Would this necessarily affect the overall electric charge? Wouldn't the particles themselves and their relative influence, also shrink accordingly? For example, if you shrink the size of an atom, wouldn't its mass also shrink? Why wouldn't it? Why wouldn't the amount of charge per particle also shrink, in a similarly relativistic way? It just does not follow logically that we should expect the overall charge to increase, such that an imbalance between negative, and positive charges is created by relative motion.

Furthermore, it is important to note that the electrons in an electromagnet are not moving in an inertial frame of reference. Rather, the electrons are moving around in a circular path. Even if it may look like there are more electrons, from a certain perspective, the actual number of electrons can be counted, and this number has not actually changed. Because you always have the same number of electrons moving around in a circle and double counting these electrons cannot be permitted, then from a more global, and comprehensive perspective, it really does *not* look like the number of electrons has increased at all. If we assume that elementary particles themselves adopt perspectives at all, then why should we assume that they adopt an incoherent local perspective, as opposed to a comprehensive, global perspective? Why should elementary particles be so easily duped? The science of working out how it is that elementary particles perceive reality, if, in fact, that is what they do, is still in its infancy. We don't and can't know how the elementary particles digest information, how they decide what to pay attention to, and what to ignore, or how they go about making decisions. Elementary particles are the simplest things that we can imagine, and so their cognitive processes are obviously very difficult to model.

The bottom line is that the total number of electrons is necessarily conserved and so is the actual ratio between the electrons and the nuclei, and this conservation must be reflected in the overall charge that is measured. Therefore, length contraction cannot explain the magnetic force. However, it is perfectly fine to accept the rules of electromagnetism as brute facts which require no deeper explanation. It is much better to have no explanation, than it is to prematurely accept a pseudo explanation.

Because magnetism had been studied long before the theory of special relativity was proposed, this necessarily means that special relativity did not predict magnetism. At best, the idea that magnetism is caused by length contraction is an awkward, post-hoc explanation of the sort that characterizes pseudoscientific reasoning.

Muons and Time Dilation

Muons are elementary particles. You can think of muons as like electrons, except that they are about 207 times as heavy, and relatively unstable. Muons only exist for extremely short periods of time before they decay.

The reason that muons often have been cited as confirming evidence of inertial time dilation, is that it has been estimated that atmospheric muons are able to cover a frame-dependent displacement of more than twenty kilometers, during their short half-life. Because muons have a half-life of only 1.56 picoseconds, then the size of this frame-dependent displacement would imply that the muons must be travelling at a relative speed that is at least five times that of the speed of light, relative to Earth's frame. Inertial time dilation is used to explain how muons can survive long enough to make this twenty kilometer trip, without moving any faster than the speed of light, relative to Earth. The idea is that the half-life of the muon must be dramatically extended, due to inertial time dilation.

The above argument undoubtedly relies on circular reasoning. Without availing ourselves of the assumption that it is impossible to exceed the speed of light, then we cannot interpret this evidence in a way that supports time dilation. We must first assume that special relativity is true, in order to construe this evidence in a way that lends support to that same theory. Alternatively, if we do not first assume that special relativity is true, then this same evidence actually falsifies, and refutes special relativity.

By way of an analogy, suppose that a relative of yours had a pet theory about the Moon and that according to this theory, the Moon sits at an altitude of twenty kilometers. Under the assumption that the moon must be twenty kilometers away, your relative can use the Moon's visual angle to calculate a lunar diameter of only 180 meters. Now that your relative "knows" that the diameter of the Moon is 180 meters, they can combine this information with the observed visual angle of 0.5 degrees, to confirm that the Moon is indeed twenty kilometers away. While circular reasoning does demonstrate a degree of internal consistency, it is not a cogent means of supporting a scientific theory because it can also be used to validate any theory, irrespective of how accurate that theory is. When a circular argument is the best argument available, this is a sign of an extraordinarily weak theory.

Simply knowing the length of the frame-dependent displacement of an object with an unknown velocity is not enough to determine either its relative velocity or the time interval. We also need to know precisely what its relative velocity is, in order to use the Lorentz equation to calculate the degree of time dilation. We cannot work backwards from the length of the path taken by the particle, in order to calculate either the velocity, or the time dilation, without making additional assumptions about its velocity. By first assuming that time dilation is occurring, in order to draw an inference about the relative speed of the muon, and then using this relative speed to conclude that time dilation is present, we would be engaging in circular reasoning. The misconception about the speed of light as a universal speed limit, which I previously classified as Hydra Head # 6, has been used to revive the entire beast.

The simplest explanation for the surprisingly long frame-dependent displacement of the muon, is that muons are capable of travelling at superluminal speeds. Hence, it would be simpler and easier to conclude that special relativity has been refuted in two ways, with just one observation. In other words, no time dilation has been observed, and the speed of light has also been exceeded.

When the velocity of atmospheric muons has been measured directly, researchers have often found that these particles do seem to be moving at faster-than-light speeds. But under the dogmatic assumption that superluminal speed is impossible, these anomalous results are normally ascribed to calibration, or measurement error, and adjusted accordingly.

Particle Accelerators and Time Dilation

Particle accelerators and colliders move extremely small and light subatomic particles, such as hadrons (protons and neutrons), at very high speeds. To observe relativistic effects, specialized synchrotron particle accelerators are used, which confine the particles to a beam that navigates a circular track. What the circular shape of the track necessarily means, is that the particles are not in inertial motion, relative to Earth. Instead of inertial motion, these particles experience accelerated motion, as they oscillate back and forth.

Incredible speeds have been achieved in synchrotron particle accelerators. The particles are apparently able to complete several thousands of laps every second, on a track which spans several kilometers. We can say that the speeds attained are very fast indeed, but because the particles are not moving in a straight line, we cannot say that inertial motion is what is being manipulated or measured. Inside a synchrotron particle accelerator, acceleration is a confounding variable, which cannot be disentangled from inertial motion.

Atomic Clocks and Time Dilation

The Infallible Atomic Clock

While a broken clock is right twice a day, an atomic clock is right always, and no matter what, by definition. In my view, it was a mistake for the International Committee for Weights and Measures to sanctify the atomic clock in 1967, by setting the definition of the second to exactly 9,192,631,770 oscillations of the caesium 133 atom. This change in the definition of the second has meant that the atomic clock has been the standard by which time is defined and measured, ever since. It is officially inconceivable that atomic clocks could be prone to error or bias.

In the Hafele-Keating experiment, atomic clocks were mounted onto supersonic jets, and flown around the world, in order to measure mere nanoseconds-worth of time dilation. It was hypothesized that travelling *against* the rotation of Earth, would result in a *gain* of time, and that travelling *with* the rotation of Earth, would result in a *loss* of time, relative to a clock that is on the ground. These assumptions raise interesting metaphysical questions which are worth exploring in more depth.

Earth's rotation is angular momentum as opposed to ordinary, inertial motion. This means that it forms a flat disk in velocity space and only the centre of this flat disk represents the overall location of Earth in velocity space. The width of the disk represents the speed of rotation, and the plane of the disk represents the plane of rotation, which is perpendicular to the axis of rotation. However, for increasingly short intervals of time, angular momentum also becomes increasingly indistinguishable from ordinary, inertial motion. Therefore, if inertial time dilation is real, then time dilation should also apply to angular momentum.

But the rotation of Earth is absolute. We can objectively determine what the absolute speed of the rotation of Earth is, as well as the axis, and direction of rotation. It is *not* equally plausible that Earth might be stationary and that the rest of the universe might actually be revolving around it (which was once the dominant view). Therefore, the angular momentum of Earth seems to be an absolute, in some sense.

But yet angular momentum is also relative, in another sense. Angular momentum is always relative to a specific axis in position space *and* to a specific point in velocity space. If we are considering the question of, to what degree Earth is rotating about its axis, and to what degree the rest of the universe is rotating about Earth's axis, then it is important to remember that this axis will be the same axis in either case, because it will have the same location and orientation in position space, and it will also have the same location in velocity space.

Through focusing on the point of contention (Earth is revolving vs. the rest of the universe is revolving), it is easy to lose sight of the agreed set of facts (the position and orientation of the axis of rotation and its location in velocity space). So, in this situation there is not really much of a disagreement at all. It is simply more parsimonious to assume that the reason that the axis of rotation just so happens to pass through the centre of Earth, and that it just so happens to match the precise location of Earth in velocity space, is because this axis of rotation is a property that belongs to Earth, specifically.

This has been a bit of a digression, but I think that it was necessary to explore how these difficult, and deep, philosophical questions apply specifically to the results of the Hafele-Keating experiment. Even though the axis and the speed of the rotation of Earth is an objective fact, and even though Earth's axis occupies an objective point in velocity space, the point that Earth's axis occupies in velocity space is still no more special than any other point in velocity space. This point in velocity space *cannot* be considered to be "motionless," in any absolute sense. Many other objects may occupy this same point in velocity space, but for different reasons, and it cannot be claimed that these objects are objectively motionless, just because they share the same location in velocity space. It still does not matter where we decide to place the zero in velocity space because all locations in velocity space are still presumably equal. Hence, it makes no sense to assume that by moving around the axis of the earth in position space, and by occupying locations that differ from Earth's axis in velocity space, one is really moving in any objective, or absolute sense. The symmetry cannot be broken so easily.

The only objective variable in the Hafele-Keating experiment arose from the fact that the plane that flies *with* Earth's rotation, covers more distance in velocity space, than the plane that flies *against* the rotation of Earth. In this case, covering distance in velocity space translates to experiencing the force of acceleration. The independent variable that was measured in the Hafele-Keating experiment was merely the degree of acceleration experienced by various atomic clocks.

The experimental hypothesis in the Hafele-Keating experiment was confirmed, but it is important to note that, in this case, the path of motion was circular, only the force of acceleration was actually being measured, and that the only standard for time measurement used, was the rate of oscillation of a caesium 133 atom.

Satellites

Time dilation effects are observed when communicating with satellites, which also rely on atomic clocks. It has been alleged that there is both a gravitational and an inertial motion-based effect that is observed when the satellite-based atomic clocks start to drift out of sync with the ground-based atomic clocks. SSL theory also predicts that some apparent inertial time dilation should be observed with satellite communications, but for a very different reason.

According to SSL theory, the signals that are received either by the satellites, or by the transmitters on the ground, are actually part of a superwave which travelled slightly faster than 299,792 km/s, relative to the sender, but at exactly 299,792 km/s, relative to the receiver. These signals must be very slightly red shifted, which gives the impression of a slightly slower oscillation rate, and a slightly slower atomic clock speed. This also means that the electromagnetic signals that the satellites receive from Earth have reached them slightly faster than we would expect them to, and the same applies to the signals received on the ground. Overall, this makes the atomic clocks on the satellites appear to be running relatively slowly, but this is mainly due to red shift, and the effect is incredibly subtle.

The claim that the atomic clocks on GPS satellites experience time dilation from inertial motion carries with it the implicit assumption that only the orbital velocity of the satellite matters, and not its direction. Multiple satellites may orbit Earth at the same speed, and at the same altitude, but in different directions. This means that in addition to moving relative to Earth, the satellites are also moving relative to one another. But the tacit assumption is that satellites which orbit at the same altitude (and therefore the same speed) will also share a common inertial frame of reference, even if they are moving in opposite directions. I have already demonstrated the problem with this line of thinking in the Twin Paradox 2.0, but I will revisit it briefly.

If the satellites are understood to be moving relative to the Earth, but not relative to each other, then it follows that their time should be dilated relative to Earth, but not relative to each other. But there is no reasonable basis for the idea that the satellites, which orbit in different directions, are not moving relative to each other.

The satellites that orbit Earth in different directions are also moving relative to each other. Therefore, if special relativity is true, then we should also see time dilation between the satellites themselves.

But because there is no asymmetry between the satellites, this also creates a paradox. There is no way for us to determine which satellites should be further ahead, and which should be further behind. If something is causing the atomic clocks to slow down at high altitudes, then it cannot be inertial motion. The total distance covered in velocity space is the only variable that is being manipulated in this case.

Common Themes in the Purported Evidence

You may have noticed a few recurring themes in the evidence that has been provided in support of special relativity. Firstly, all of these purported examples of time dilation and length contraction rely solely on observations of quantum objects. We do not have any confirming observations with respect to classical objects, such as golf balls, planets, or people (classical objects are a very broad category). Secondly, with the notable exception of muons, all of these effects depend on accelerated, and/or circular motion, as opposed to linear, and/or inertial motion. In my view, it is not reasonable to generalize from the observed effects of acceleration on quantum objects, to the effects of inertial motion on classical objects. There is no good reason to expect that classical objects should behave exactly as quantum objects do at high relative speeds, but there are many good reasons to expect that classical objects should behave differently from quantum objects at high relative speeds. Ideally, it would be necessary to perform controlled experiments, in order to determine how classical objects respond to extreme acceleration, how quantum objects respond to extreme inertial motion, and how classical objects respond to extreme inertial motion.

I have represented what the empirical evidence has demonstrated to date, in the chart below. The box with the check mark indicates that there is empirical evidence of time dilation for quantum objects, which are in circular, and/or accelerated motion. But in order to truly confirm the predictions of special relativity, all four of these boxes would need to contain check marks. The ultimate confirmation would be an observation of inertial time dilation, with respect to classical objects. The other two boxes represent the intermediate steps, which have also not been tested, or confirmed yet.

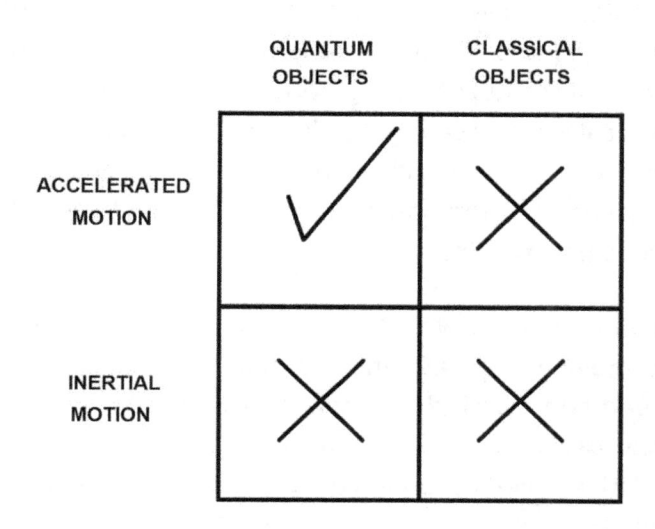

	QUANTUM OBJECTS	CLASSICAL OBJECTS
ACCELERATED MOTION	✓	✗
INERTIAL MOTION	✗	✗

What caesium 133 atoms, muons, hadrons, and electrons all have in common, is that these are quantum objects. The behaviour of quantum objects is radically different from that of classical objects because they obey the laws of quantum mechanics as opposed to the laws of classical Newtonian mechanics. The vast gulf between the behavior of quantum mechanical objects and classical objects is universally recognized, irrespective of whether the Copenhagen, many-worlds, or pilot wave interpretations of quantum mechanics are favoured.

Briefly, it is neither the size nor the mass of an object, per se, that gives it a special quantum mechanical character. Quantum mechanical properties appear to result, more directly, from the degree of isolation of an object, from its environment. A classical object is well-integrated with its environment, in a way that a quantum object is not. This integration is known as *entanglement*, in the quantum mechanical lexicon. Due to their isolation, quantum mechanical objects retain a number of freedoms that classical objects do not have. Even though we label these quantum mechanical objects as particles, it is more useful to think of them as waves. In other words, out of necessity we have been systematically conflating a quantum mechanical effect with a relativistic effect. While the distinction between classical and quantum objects is really more of a continuum than a binary category, treating this distinction as a binary category works adequately for our present purposes. In my second book, I will dive much deeper into the nature of the division between classical objects and quantum objects.

It is easy to understand why it is that we have never been able to experimentally manipulate and examine the effects of extreme relative speeds on classical objects. We simply do not have the technological capacity to accelerate a classical object, such as a bullet, to relative speeds in excess of 100,000 km/s. By comparison, a typical bullet moves at a relative speed of less than one km/s.

The other common theme amongst the various examples, is that inertial motion is never held constant, and manipulated independently from acceleration. The original research question was about how relative inertial motion affects the relative rate of the passage of time for classical objects and classical observers, such that the observed speed of light is held constant. Unfortunately, the electrons, hadrons, and caesium atoms all travel along a curved path, and are subjected to constant acceleration. Perhaps applying a force of acceleration and/or gravitation has an effect on the rate of oscillation of these wavelike quantum mechanical objects, which is then mistakenly interpreted as a relativistic effect, caused by their inertial motion. Not only can this quantum mechanical interpretation not be ruled out, but it also seems to be the most plausible interpretation that is available.

Predictions of the Supervelocity and Selection Theory

The Basics

Imagine that an alien spacecraft from the Andromeda galaxy is racing towards Earth at half of the speed of light. This spacecraft also has a clock on board, and with each passing second, it sends out a flash of light which is visible to an observer on Earth. What is the time interval between flashes that the observer on Earth will measure? Special relativity says that the observed time interval will be influenced by both time dilation and the Doppler effect. SSL theory says that the observed time interval will depend only on the Doppler effect. In the chart below, I compare the predicted, observed time intervals that are supplied by each theory, and under four different scenarios. In these four scenarios, the spacecraft either approaches, or recedes from Earth, and it does so at either ninety percent or fifty percent of the speed of light.

Predicted Time Intervals

Velocity	SR	SSL
Towards at 0.9c	0.22942	0.1
Towards at 0.5c	0.57735	0.5
Away at 0.5c	1.73205	1.5
Away at 0.9c	4.35890	1.9

It is well beyond our current technological capabilities to test or verify any of these predictions. However, we may soon be able to send a spacecraft hurdling away from Earth, at more than ten percent of the speed of light, while emitting a timed signal at regular intervals, and this could make it possible to put these competing theories to the test.

Time Intervals depend on Locations in Velocity Space

Let's imagine that a flash of light is emitted from a source and that there is an observer who we will call Observer A, who maintains a fixed distance away from the source of one light year. Because Observer A is not moving relative to the source of light, they will see the flash exactly one year after it has been emitted.

Now, let's imagine a second observer, who we will call Observer B, who is moving away from the source at half of the speed of light. At precisely the same moment that the flash of light is observed by Observer A, Observer B passes right next to Observer A, such that both observers are exactly one light year away from the source, at that moment, which is also one year after the light was emitted from the source. But even though Observer B is at the same location as Observer A at the moment when Observer A sees the flash, Observer B does not see any flash of light, at that time. This result is highly counterintuitive, but it follows from SSL theory.

Observer B would still have seen the same flash of light, but they would have seen it only six months after the flash was emitted, and at the time when they were only 0.75 light years away from the source. In other words, Observer B would see the flash of light before Observer A sees it, and before they pass by Observer A's location.

I should point out that the results of this thought experiment are fully consistent with the two core postulates of special relativity. According to these postulates, the light that is seen must travel at 299,792 km/s, relative to each observer, but it can also travel at any speed, relative to the source. From these principles, it does follow that Observer A, and Observer B, must see the same flash of light at different times, despite being the same distance away from the source, at the same moment in time. It is impossible for both observers to see the same flash of light at the same time. It is apparent that the same flash of light has travelled at two different speeds simultaneously, one speed for each observer.

If we also have an Observer C, who is moving rapidly towards the source at half of the speed of light, then Observer C would be the last observer to see the flash, as the opposite effect would manifest. Observer C would also see the flash when they are 0.75 light years away from the source, but they won't see the flash until eighteen months after the flash was emitted.

Intergalactic Oscillations in Time and Pulsars

One implication of the SSL theory is that the apparent age of distant galaxies must fluctuate on an annual basis, from the perspective of observers on Earth. The cyclical change in the orbital velocity of Earth around the Sun would be responsible for this hypothesised effect. As Earth moves relative to the Sun, in the direction of a distant galaxy (towards), that galaxy should appear to be younger because we would be able to see further back in time. While Earth is moving in the opposite direction (away), then the galaxy should appear to be much older, as we will be looking back a shorter distance in time. By contrast, special relativity does not predict this effect because according to special relativity, we are observing the same light all year round, and it is the relative rate at which time passes for Earth that changes only very slightly, due to variations in our relative motion. According to special relativity, the observed effect of time dilation would not differ, depending on how far away the distant galaxy is, or on how long it has taken for its light to reach us. But according to SSL theory, the distance of the galaxy matters a great deal. All else being equal, the further away the distant galaxy is, the more pronounced this observed temporal oscillation effect will be.

The Andromeda galaxy is the closest galaxy to the Milky Way, and it is the easiest to observe. According to current estimates, the

Andromeda galaxy is roughly 2.5 million light years away from the Milky Way. Therefore, if both galaxies were in the same location in velocity space, then it would take the light from Andromeda 2.5 million years to reach the Milky Way. But if the light that we see from Andromeda were to travel a mere 0.01 percent faster than c, relative to Andromeda, then it will take 250 fewer years for that light to reach the Milky Way. Therefore, the apparent age of the Andromeda galaxy should vary by about 500 years, depending on the time of year at which it is viewed at from Earth.

To say that a typical galaxy does not appear to change much over a period of 500 years is an understatement. Because the Andromeda galaxy takes hundreds of millions of years to complete a single rotation, a period of 500 years will only result in about one 1,400th of a degree of rotation, or 0.0007 degrees of rotation. Such a small difference will be extremely difficult to detect, but not impossible, in principle. With enough optical resolution power and determination, it should be possible to watch the Andromeda Galaxy rotate both forwards and backwards, without relying on proxies such as red shift to estimate the rotation.

In other words, due to the annual change in Earth's velocity, relative to the Andromeda galaxy, we should see the apparent age of the galaxy fluctuate back and forth by roughly 500 years, each year. This also means that as Earth approaches its furthest position from Andromeda each year, we could actually watch the Andromeda galaxy move backwards in time and rotate backwards. As we reach our closest point in our orbit, we could be watching the time on Andromeda move in hyper-fast motion. There would also be a special moment, twice every year, when time on Andromeda appears to stop, as our acceleration, relative to the Andromeda Galaxy, is near a minimum.

One minor caveat, is that the Andromeda galaxy is also approaching our solar system at roughly 300 km/s. This caveat does not significantly affect the calculations or the conclusions that I have reached, or the implications for the testing of the theory.

A second caveat is the fact that the location of the Andromeda galaxy is not perfectly aligned with the ecliptic of Earth. The angle of the Andromeda galaxy, relative to the ecliptic plane (the imaginary plane which contains Earth's orbit around the sun), will need to be considered, and this means that the effect of Earth's orbit will be slightly lower, and more difficult to detect. Because the Andromeda galaxy is roughly thirty degrees away from the ecliptic plane, then we must multiply the degree of the effect by the cosine of 30 degrees, or 0.87. Hence, rather than a variability in age of 500 years, it should be closer to 440 years. As these numbers are all very approximate, these small differences are hardly significant.

A third caveat is the fact that the Andromeda galaxy is more than 100,000 light years across. Therefore, the light that we see from that galaxy did not all leave from the galaxy at the same time. The light from closer parts of the galaxy left tens of thousands of years after the light that we see from the more distant parts of the galaxy. This large range of tens of thousands of years makes it impossible to see what the entire galaxy looked like, at any particular moment in time. This complication will make it significantly more difficult to observe a relatively small oscillation in time, of a mere 400, or 500 years.

Another potential way to test SSL theory would be to look for and study pulsars in the Andromeda galaxy. If SSL theory is false, then any pulsars that we are able to identify in the Andromeda galaxy should display a regularly timed pulse, year round. If SSL theory is true, then we will be able to observe a marked decline in the pulse rate of these stars, twice every year. It is difficult enough to detect pulsars within our own Milky Way galaxy, and so it will be very challenging to reliably identify intergalactic pulsars and monitor them closely enough to measure their pulse. But this is not impossible, and some recent progress has already been made, on this front. We already know that pulsars almost certainly do exist in the Andromeda galaxy, and so all that is left to do is to locate just one pulsar and to observe it closely.

We should also be able to detect the temporal oscillations that are predicted by SSL theory, by looking within our own galaxy. The best candidates for stars which we could observe for the purpose of looking for temporal oscillations would be pulsars and type I Cepheid variable stars. The ideal stars would be those stars for which we have obtained an exceptionally high degree of confidence and precision, with respect to their distance from Earth. The best candidates would also be located close to the ecliptic plane and somewhere between 400 and 4,000 light years away from Earth.

According to SSL theory, there is a very special circle which is contained within the ecliptic plane of Earth and which extends roughly 1,600 light years outwards, in all directions. This circle marks the boundary where the maximum slope of the predicted time oscillation is equal to the rate at which time actually passes for that object. This means that at one special point during the year, any object that is viewed from this special distance, should appear to be briefly frozen in time. This special moment would occur at the point when Earth is furthest away from the distant object. For visible objects that occupy a distance that is significantly further away than 1,600 light years, it will also be possible to see them as if their time were running backwards, provided that they are at, or near the ecliptic plane. To put this distance of 1,600 light years into perspective, the closest stars are a mere four light years away from Earth, and even these relatively nearby stars are roughly 300,000 times as far away from Earth, as compared to the Sun.

The degree of observable temporal oscillation would depend on both the angle of declination of the star and the distance to the star. A pulsating star that is one thousand light years away would appear to speed up and slow down its pulse rate much more dramatically than a pulsating star that is only one hundred light years away. A star that is at a declination of fifteen degrees would also speed up and slow down more dramatically, than a star that is at a declination of seventy-five degrees. Through making observations of the same type I Cepheid variable star, or pulsar, at different times of the year, it should be possible to observe the predicted annual changes in their pulse rates.

Despite the caveats that I have mentioned, SSL theory does make a few clear predictions, and it is falsifiable. Through taking advantage of the parallax in velocity space that is afforded by our planetary orbit, SSL theory predicts that we should be able to detect significant temporal oscillations at great distances.

Rapidly Moving Mirrors

What does SSL theory say about the one-way speed of light?

SSL theory says that light doesn't travel at any particular speed, but rather it travels at a range of velocities, and speeds, simultaneously. There is no known upper limit to the speed that light travels at, and this is why its speed manifests, in some ways, as infinite. One might think that the lower limit on the speed of light would be zero, but this is false. The lower limit would be the negative reflection of the upper limit. Because negative speed is just positive speed, but in the opposite direction, this explains why light tends to radiate symmetrically, in all directions, at the same time.

In other words, rather than thinking about light that moves in the opposite direction as different light moving in a different direction, it is better to think about this light as being the same light, and as moving in the same direction, but at a much lower speed.

According to SSL theory, the manifest light that interacts with the observer travels at roughly 299,792 km/s, relative to the observer, and directly towards the observer. If a mirror is in the same location in velocity space as the observer and the light that the observer sees is reflected by the mirror, then that light would also be moving at 299,792 km/s, relative to the observer, before it was reflected.

However, everything changes when the observer and the mirror are not in the same location in velocity space. In that case, the manifest light that the observer interacts with still travels at 299,792 km/s, relative to the observer, but it travels at a different velocity, relative to the mirror. If the mirror is moving away from the observer, then the manifest light is moving faster than 299,792 km/s, relative to the mirror. The corresponding incident light is also moving at this faster speed, relative to the mirror, and even faster, relative to the observer.

I am making two critical assumptions here: 1. That a mirror reflects latent light just as it reflects manifest light. 2. That the latent light which is reflected by the mirror would be reflected back at the same relative velocity as the incident latent light.

If the mirror is moving away from the observer at 0.5 **c**, then the manifest light that reflects off of the mirror moves at 1.5 **c**, relative to the mirror. Therefore, this light was moving at 2.0 **c**, relative to the observer, before it was reflected off of the mirror, and so it could not have been seen by the observer.

If the observer sees one pulse per second, then this means that the moving mirror was reflecting the pulses at a rate of 1.5 pulses per second. The source would have been releasing the pulses at a rate of two pulses per second. Hence, there is a strong red shift that should be observed in this case, as the pulse rate has apparently decreased to half of what it was initially.

Given that the speed of the manifest light is moving at 2.0 **c**, relative to the observer, for half of the trip to the mirror, and back, then this should result in a decrease in the amount of time that it takes for the observer to see the reflected light. Both the decrease in the delay between the emission and the observation of the light and the red shift, are potentially empirically observable consequences of SSL theory.

Interpreting Red Shift with SSL Theory

Let's say that the distance between the Milky Way and some distant galaxy is growing at a rate of 150,000 km/s. According to the SSL theory, the light that we see coming from that galaxy travelled at 450,000 km/s, relative to that galaxy, but only at 300,000 km/s, relative to us. Because that light would have travelled at 450,000 km/s, relative to that galaxy, it would have also been totally invisible to anyone who inhabits that galaxy. The wavelength of that light would also have to have been 1.5 times as long, as compared to the light that can be seen within that galaxy. The wavelengths of the light that we see from that galaxy must, therefore, also be 1.5 times as long as what we would otherwise expect to see. In other words, galaxies that are moving away from us must also appear to be red shifted, and this is precisely what we do see. We have also derived and predicted red shift, but in a different way from how this is normally done in astrophysics.

I say that the SSL method is different because it relies on a different set of assumptions. One important, underlying assumption is that the frequency of the electromagnetic radiation that emanates from a source remains fixed, and so as the speed of the wave relative to the source increases, then so does its wavelength. A superposition of speeds entails a superposition of wavelengths, as well. Faster speeds emanating from a source entail longer wavelengths. Slower speeds emanating from a source entail shorter wavelengths. Reversed or negative speeds emanating from a source entail reversed or negative wavelengths.

Conservation of Energy and SSL Theory

Just as a superposition of photons and electrons does not violate conservation of energy, neither does a supervelocity of photons. Both sides of the energy ledger will still balance under SSL theory. No energy will be lost, and no energy will be created. The overall picture, with respect to the conservation of momentum, and the conservation of energy, will look the same as it did before.

I mention this, because the sudden discovery, and acknowledgement of a vast amount of latent light, may give some the impression that there is extra energy being released that needs to be accounted for. However, both the total release of energy from light sources and the total absorption of energy remain unchanged under SSL theory.

Let's imagine that the energy of ten photons is released from a source. This means that a maximum of ten and a minimum of zero photons will be observed. If all of these photons land on the same screen then their individual positions on the screen will be randomly scattered. If there are twenty screens at different distances from the source then only a maximum of ten screens will detect at least one photon. Both the screens and the positions of the photons on the screens would be randomly selected. If these twenty screens are moving at different velocities relative to the source then only a maximum of ten screens will detect a photon. You will *not* end up with twenty times as many photons from the source. The more screens you have, the lower the probability of detecting a photon will be per screen. The total number of photons is conserved regardless of whether the screens are in different locations in ordinary space, or in different locations in velocity space.

Important Questions

What is the Standard Deviation of Manifest Light?

Manifest Light Must Have a Gaussian Distribution

The speed of the light that we see is not precisely and exactly 299,792,458 m/s, relative to the observer, because this is only an average value. The speed of the manifest light is actually a Gaussian distribution of values, and 299 792 458 m/s is close to the midpoint of this Gaussian distribution. Any prior illusions about obtaining absolutely perfect precision and accuracy should have been shattered through the study of either quantum mechanics, chaos theory, or signal detection theory. In particular, the Heisenberg uncertainty principle tells us that there is always some minimum degree of blurriness, with respect to the precise position, or momentum, of any object. Therefore, there should be little doubt that the manifest speed of light does conform to a Gaussian distribution, of some description.

The only real question with respect to the Gaussian distribution of the manifest speed of light, is how large its standard deviation is. This means that the selection criterion is a range of values as opposed to a single value, and the question is about how large this range is. To use the analogy of a pupil, we know that a pupil is very small, but it is also larger than a single point with zero area. So how large is the pupil that we use to see light in velocity space?

Absolute Precision by Definition

Before I continue, I should address one minor caveat. In 1983, the definition of the meter was changed by the International System of Units, from a stable physical standard to the distance that light travels in a vacuum in 1/299,792,458 of a second. In other words, it is now true by definition that the speed of light is exactly 299,792,458 m/s, in a vacuum. This is not a rounded number. This is not an estimate. There is no conceivable possibility for error. This precise value has not been empirically derived from or confirmed by any experimental results. This number was actually derived from a mere consensus, and therefore it is inherently arbitrary. No experiment could possibly demonstrate that the speed of light has any value that deviates from the officially recognized value because no matter what the experimental results turn out to be, it is impossible to either validate or refute a claim that is simply true by definition.

Hypothetically, if a well-conducted experiment was to correctly demonstrate that light had in fact travelled 301,458,902.1371 meters in one second, then all that this would mean is that what the experimenters had initially measured accurately, as a distance of 301,458,902.1371 meters, would be retroactively redefined as 299,792,458 meters. This is because the definition of a meter is no longer based on any stable, physical distance. Hence, if the experimentally observed speed of light changes, then the definition of the meter must also change accordingly, such that the officially measured speed of light will wind up being exactly the same as it was before. Progress is pre-emptively precluded by the current definition. It would make no difference how the scientists actually measured the distance of the path that the light travelled, because the speed of light is now officially recognized as the only standard by which distances can be measured, and calibrated.

I call this a "minor" caveat because I assume that it should be universally acknowledged that the behaviour of light and the laws of physics do not change, depending on how human institutions elect to define their units of measurement. To declare that something is "true by definition," in this context, is to construct a tautology, and therefore it cannot have any bearing on either the empirical evidence or the immutable laws of physics. We must be very careful not to find ourselves trapped in circular reasoning, and the redefinition of the meter in terms of the speed of light creates a potential hazard for circular reasoning. It would be inappropriate and absurd to allow this arbitrary and stultifying definition of distance to dissuade us from investigating the fine details of the real physical phenomena.

How Thick is the Membrane of the Light Bubble?

Another way to look at the question about the standard deviation of manifest light, is to ask about the dimensions of the corresponding light bubble, in velocity space. Because the location of the observer in velocity space cannot be determined with absolutely perfect precision and accuracy, then neither can the location of the light bubble which surrounds the observer. The thickness of the membrane of the light bubble that surrounds the observer is the standard deviation of the relative speed of the manifest light.

According to SSL theory, latent light travels in a superposition, in terms of its speed, its direction, and its wavelength. Faster speeds correspond to longer wavelengths, and slower speeds correspond to shorter wavelengths. Any measurable variability in wavelength would also correspond to variability in speed. Measuring the thickness of the light bubble can therefore be accomplished by measuring the variability in wavelength that is associated with light that is as monochromatic as possible. One can think about this variability in wavelength as akin to a randomized Doppler effect, that is not caused by relative motion.

Care must be taken to avoid any averaging while measuring this value, as this would cause the true variability to be underestimated. The fundamental range in the energy of individual photons from a monochromatic light source is probably the most promising way to gauge this variable, but it should also be possible to converge on a solution from multiple lines of evidence. If averaging is used for one reason or another, then the effect of this averaging on the standard deviation must also be considered. It should not be assumed that all observed variance is necessarily due to measurement error. Instead, it will be necessary to assume that at least some of the variability that we find, is variability in the relative velocity of manifest light itself.

The empirical evidence that has been collected to date implies that this variability is less than one in 1,000 but more than one in 10,000. The most precise measurement of the smallest spectral frequency band, which has a wavelength of 21.106 cm, has been narrowed down to a range of 1420.4 MHz to 1420.7 MHz. This variability in wavelength corresponds to a variability in the speed of light of 299,789,624 m/s, at the low end, to 299,852,942 m/s, at the high end.

Other related questions that come to mind are, why it is that our methods for detecting the presence of electromagnetic waves are only capable of interacting with waves that move at a particular speed, relative to the measurement device? What is it that all electromagnetic waves that move at a particular relative speed have in common? Why is it that all of the waves that move at speeds outside of this narrow range are able to pass by us undetected? Is it possible that we have detected electromagnetic waves that move at different speeds without recognizing them?

One intriguing possibility that I take seriously, is the idea that the light waves that we see are actually akin to moiré patterns caused by aliasing, which would result from the interaction, or interference between multiple lattices. At least one of these lattices would be grounded in the observer's location in velocity space. I haven't given this possibility a great deal of thought, but I also haven't yet found a compelling reason to rule it out.

What Is the Size and Shape of the Supervelocity?

What is the size of the entire distribution of latent light from a given source? What is the shape of this distribution? Is the distribution evenly spread out over velocity space? Is the size of the distribution finite? Is the size of the distribution variable? It is unlikely that an empirical investigation will be able to answer any of these questions in the near future. All that we can know is that the size of the distribution of the latent light must be significantly larger than that of the manifest light. The actual ratio between the size of the distribution of latent light and manifest light might be one to one thousand, one to one million, or one to one billion. So far, we do not have any way to conclusively rule any of these possibilities out.

Can we Shed Light on the Enigma of 137?

It is difficult to overstate the well-deserved reverence and significance that the number 137 has earned in theoretical physics. If humanity had greater interest in theoretical physics, then we would be erecting monuments and shrines to honour the number 137. The inverse of 137 is also known as the fine structure constant or alpha. I strongly suspect that defining the relationship between manifest and latent light may finally explain this captivating and perplexing number.

The fine structure constant was discovered pursuant to the discovery of the Lamb shift, which is a small anomaly in the wavelengths of the light that is emitted by hydrogen atoms. This anomaly had been difficult to explain, but it was eventually understood with the help of the fine structure constant.

137 is also the ratio of the speed of light compared to the speed of an electron in a hydrogen atom. The photon is thought to be 137 times as fast as the electron. This is an interesting conclusion, given that according to quantum mechanics, an electron is in superposition

while bound to an atom. It is strange to think of the electron as moving at any particular speed because an electron should be everywhere that it can be and all at the same time. Therefore, what we interpret as momentum for an electron is qualitatively different from the sort of momentum that we can measure for classical objects. Nonetheless, this special number does tell us something important. 137 also relates the magnitude of virtual particle interactions to their complexity in Feynman diagrams. Hence, this number is used to relate latent interactions to manifest interactions.

It may be that 137 can be explained as a product of the ratio between manifest and latent light. This number might reflect the size of our metaphorical pupil and our distance from the source in velocity space. If I am standing a certain distance away from a point light source, then the percentage of the light that I can see drops off, as the finite area covered by my pupil shrinks, relative to the area of a sphere, with a radius of my distance from the source. In other words, what we are trying to determine here, is the visual angle of the metaphorical pupil of an observer in velocity space. What is the maximum percentage of latent light that can become manifest light for a particular observer?

Is the Supervelocity Potential or Actual?

In theoretical quantum mechanics, an abstract quantum superposition or wave function may be regarded either as a mere distribution of potential positions, or as a real object in its own right. There is a similar question in theoretical quantum mechanics, as to just how real so-called "virtual particles" are. Along the same lines, we can regard a supervelocity merely as a distribution of potential velocities, or we can regard every velocity that exists in that distribution, as actual but unobserved. Another way to put this question, is to ask whether or not velocity space is a type of "phase space," similar to that employed by quantum mechanics. Just as with theoretical quantum mechanics, it is best to leave this question open for as long as possible because it is difficult to reach conclusions as to how real these potential and unobservable entities are. There may also be indirect means of observing them, such as through the generation of interference patterns. Indirect methods of observation might indicate that they exist as something more coherent and stable than as a mere abstract potential.

Further Questions about the Rainbow

The title of my book is a question about the rainbow. A valid answer to that question might be something like "everywhere, but nowhere in particular." And this type of answer would also be valid as a reply to the analogous question about the speed of light. Contrary to popular wisdom, the speed of light is everywhere, but nowhere in particular. But there are several other basic questions about the rainbow which I do not yet have answers for.

I find it fascinating that when light travels through a medium, it also changes in three important ways. Light appears to change its speed, its direction, and its wavelength. Only the frequency of the light remains constant. Even more fascinating, the degree to which the wavelength of light changes, itself depends on the original wavelength of the light. This opens up the possibility that there is some special, long wavelength of light which remains the same, or that there is some special, short wavelength that will be compressed into nothing. This naturally leads me to ask deeper questions about the nature of refraction.

One might expect that there would be some scientific consensus about why and how the speed of light slows down at the moment when it enters a refractive medium. But there is no scientific consensus around these questions. Instead, there is only a morass of competing theories. The most popular theory also appears to be the least plausible. This theory says that individual photons are briefly absorbed by individual atoms and then re-emitted, at a later time.

In other words, there is a slight delay in the relay of light between individual atoms.

According to SSL theory, refraction does not change these three properties of latent light at all. Rather, the manifest light has temporarily shifted from one speed and wavelength to another. For some unknown reason, this shift is more dramatic for shorter wavelengths of light.

The following questions have the potential to lead to new clues about the nature of light:

1. Why is it that shorter wavelengths are refracted more than longer wavelengths when they enter a medium such as water?

2. Why is it that shorter wavelengths are slowed down more than longer wavelengths, as they traverse a medium?

3. Why does the optical density of the medium affect shorter wavelengths more than it affects longer wavelengths?

4. What are the shortest and longest possible wavelengths of light?

5. Are there any real limits on the possible wavelengths of light?

6. When light traverses a medium, does its velocity become fixed relative to that medium, or does its velocity remain observer-centred, in the form of a superwave?

7. Is there any simple, mathematical relationship that can be used to calculate the observed index of refraction, based on the interaction between the optical density of the medium, and the wavelength of the light? If so, then such a mathematical relationship would provide us with a wealth of insight and possible clues that could lead to answers for most

of the unanswered questions that I have, about the creation of the electromagnetic spectrum.

There is some ambiguity and confusion with respect to how the concept of a refractive index is understood. All the refractive index does, is mathematically relate the angle of incidence to the angle of refraction, such that if you have one, you can calculate the other. The relationship between these angles and the index of refraction, is given by Snell's law. However, a refractive index may be understood, either as a general property of a medium which depends on the optical density of the medium, or as a property of how a medium interacts with a specific wavelength of light.

For example, the refractive index of water is recognized as roughly 1.33. However, this refractive index of 1.33 really only applies to yellow light, which is in middle of the visible part of the electromagnetic spectrum. For the purposes of the chart below, I have construed the index of refraction as a general property of the optical density of the medium, and I have calibrated this measure to light with a wavelength of 592 nanometers.

If we fix the angle of incidence to precisely sixty degrees from normal, if we have a range of wavelengths from thirty-seven nanometers to 37,888 nanometers on the **x** axis, and if we have mediums with various optical densities, as represented by an index of refraction that is calibrated to 592 nanometers on the **y** axis, then the chart below depicts how the angle of refraction varies, depending on both the wavelength, and the optical density of the medium.

Index of refraction \ λ (nm)	37	74	148	296	592	1,184	2,368	4,736	9,472	18,944	37,888
1.25					43.85						
1.33			34.68	39.55	40.52	40.89					
1.50					35.26						
1.75					29.66						
2.00					25.66						

The reason that the chart above is almost entirely blank, is that most of the data required to complete it is either nonexistent, or very difficult to find. There may or may not be a good reason for this. For example, what the introductory textbooks have to say about the relationships between these variables may be vastly oversimplified, and so there may not be any recognizable pattern to be found in such an analysis. Perhaps, the index of refraction does not adequately capture the full complexity of how light interacts with a given medium. It might also be the case that the clear patterns we see in the visible range (370-710 nanometers), do not normally extend to other parts of the electromagnetic spectrum. I have used a combination of Snell's law and the medium specific index of refraction (which is calibrated to yellow light) in order to fill out the visible portion of the chart. I was also able to find a recent study that looked at the refraction of ultraviolet and infrared light, in distilled water.

I had once naively hoped that there was some master equation which could relate all of these variables and could provide the angle of refraction if given the angle of incidence, the optical density of the medium, and the wavelength of the light. If the reader wants to attempt to derive such an equation or to fill out the chart above with the relevant data, then I would appreciate the effort.

In conclusion, special relativity is not as coherent or as scientific as is commonly believed. In the absence of any viable alternative to special relativity, physicists have been inclined to overlook its inadequacies. Special relativity also seems to follow logically from its core postulates, but only if we first assume that light has a single, defined velocity. This assumption is both natural and understandable, especially given that light normally does appear to have a single, defined velocity. But if we consider the possibility that light can travel at multiple velocities at the same time, then we find that the bulk of special relativity does not actually follow from its two core postulates. The Supervelocity and Selection theory of Light can solve the same problems that special relativity has been tasked with, but much more efficiently. The use of velocity space also makes it much easier to see both the relevant problems and their solutions. However, the Supervelocity and Selection theory of Light does raise more questions than it answers, and it still needs to be empirically evaluated.

To answer the question posed by the title of the book ... because the rainbow only exists as a two-dimensional image on the back of the observer's retina, the rainbow really is in the eye of the beholder.

Preview of Book Two

If you have appreciated this book then you may be pleased to know that I also have a second, much longer, and far more ambitious project, that is nearly complete. While this book focuses on relativity, the second book will be focus on quantum mechanics. Although it is relatively difficult to summarize the overall thesis of the second book, I have managed to distill its main conclusions down to the following eleven points:

Understanding Entropy as Spherical Wave Dynamics

I propose that the second law of thermodynamics is actually a consequence of spherical geometry and the predictable dynamics of radiation. Entropy must increase over time *because* waves and radiation always radiate outwards, rather than inwards, as expanding concentric spheres of energy. This geometrical constraint is the mechanism behind the "arrow of time" aspect of entropy and thermodynamics. Forwards in time corresponds to radiative expansion and backwards in time corresponds to radiative contraction. The second law of thermodynamics is therefore deterministic, and not probabilistic as currently conceived. The probability that entropy will decrease is exactly zero. The established explanations for the second law of thermodynamics, which have been supplied by a combination of statistical mechanics, and information theory, are therefore redundant and should be discarded.

In addition, I will demonstrate that Boltzmann's definition and equation for entropy vastly underestimates complexity by regarding all particles of the same type as equivalent and interchangeable. Because particles are regarded as mere tokens, or symbols, from the outset, this leads to an understanding of matter as being reducible to information.

Probability as Subjective, Abstract, and Dynamic

I will make the case that probability is fundamentally subjective, abstract, and dynamic. These three aspects of probability are very easy to demonstrate but also very easy to forget and neglect. Currently, quantum mechanics treats probability as if it were objective, concrete, and static. Representing a probability as a single number contributes to the misleading tendency to oversimplify it. Our use of second order probabilities also reinforces this tendency. The upshot of my argument is that if the wave-function describes a probability, then it is describing something that is subjective, and abstract. But if the wave-function describes something that is objective and concrete, then what it is describing must be something other than a mere probability.

Information is Created and Destroyed but Not Conserved

Contrary to the consensus in theoretical physics, information is not conserved. I will explore the nature and origin of information and I will conclude that information is inextricably tied to information systems and/or living systems. Information is ultimately defined by its functions, which include replication, representation, communication, perception, memory, and cognition. To distinguish real information from pseudo-information, I apply the term *Shannon information*, and I will show that it is inappropriate to assume that the rules of Shannon information also apply to pseudo-information. Defining the boundary between real information, and everything else, also provides insight into solving the infamous *measurement problem*. Because Shannon information is not conserved, black holes, and the big bang do not present us with any information paradoxes.

Conflating Information: Shannon, Laplacian, and Quantum

What theoretical physicists refer to as *information*, actually consists of three radically distinct concepts. Shannon information, Laplacian information, and quantum information have hardly anything substantial in common with one another. However, these three distinct meanings of *information* are frequently conflated and treated as equivalent and interchangeable by theoretical physics. Laplacian information appears to be the common core concept which serves as a bridge to unite the other two. Confusion between different meanings of the term *information* leads to a large number of significant errors and misconceptions. I will unpack the various misconceptions that underpin established physical concepts such as the Landauer limit, the Bekenstein bound, von Neumann entropy, information entropy, and the holographic principle.

Characterizing Decision-Making Within Compatibilism

Free will, agency, and decision-making are fully compatible with determinism. This compatibilist position is controversial because it tends to separate the philosophers from the physicists, but it also carries implications for how decision-making is conceptualized. Free will and agency should be seen as a continuous dimension rather than as a binary category. I identify three objective factors that contribute to free will and agency. 1. Internal vs. external mechanisms for decision-making. 2. Causal divergence vs. causal convergence in the environment. 3. Goal-directed vs. arbitrary decision-making. For free will you need a combination of internal mechanism, causal divergence, and goal-directedness. There is sufficient room for relative free will and agency within a fully deterministic universe.

Retro-Causality and Decision-Making

The primary locus of causality rests on decision-making. While causes normally do precede effects in time, there may also be limited exceptions to this rule, and the result would be retro-causality. It is decision-making which ultimately separates causation from mere correlation. Correctly sorting out causal relationships is necessary, when grappling with questions about locality, and realism in quantum entanglement. If we have no capacity to directly manipulate a variable, then we cannot draw inferences about the causal direction behind the observed correlations. While the violation of Bell's inequality is normally interpreted as a contest between the principles of locality and realism, I say that neither locality, nor realism, holds. My causally agnostic interpretation of the EPR paradox is much more cautious than the standard interpretations are, but this also opens the door to more intriguing possibilities.

Integrating Chaos Theory with Quantum Mechanics

Moving forwards in time is equivalent to magnification or amplification in space, with respect to increasing resolution, and clarity. Amplification and damping are also two sides of the same coin. One consequence of this special relationship between time and space is that Laplacian epistemological determinism is refuted. A second consequence is that Laplacian information must both be created and destroyed. A third consequence is a long overdue unification between chaos theory and quantum mechanics.

Replacing Information Theory with Signal Detection Theory

Information theory must be replaced with signal detection theory, in order to comprehensively interpret the manifestation of randomness in quantum mechanics. What may appear to be information entropy or information compression, is really just noise. The universe does not operate according to the same principles as an information system, and so the application of information entropy to quantum thermodynamics is a mistake.

On the Properties Ascribed to Elementary Particles

I claim that individual elementary particles do not really possess the macro-properties that are routinely ascribed to them, such as angular momentum, polarization, position, and linear momentum. What is ordinarily interpreted as the property of an elementary particle is often just a consequence of a larger macro-property of a macro-system. The information that we collect about an individual particle tells us nothing about that individual particle. Rather, this information is raw data, which upon aggregation, only gives us information about the properties of the relevant macro-system. This is a pattern that repeats across many different instances of quantum mechanical phenomena.

Quantization, Thresholds, Arbitrariness, and Randomness

I will be describing the relationship between hyper-constraint, arbitrariness, randomness, quantization, and thresholds between continuous, and discrete variables. I will argue that the paradoxes presented by hyper-constraint create the necessity for under-determination. Under-determination also causes arbitrariness, and arbitrariness causes what we observe as quantum randomness. The existence of a dynamic transition between continuous and discrete variables and the thresholds which are found at the boundary between continuous and discrete variables, also play an important role in both quantization, and randomness.

The Inherent Inversion of Quantum Mechanical Representations

The relationship between the representation and the corresponding reality is inverted in quantum mechanical experiments because the representation becomes more complex than, and both spatially, and temporally larger than the reality that it represents. It takes an incredible amount of knowledge, sophistication, and effort to successfully invert the natural relationship between the representation and the reality. The implications of this exotic, fragile, and extreme inversion are ultimately behind the observed quantum weirdness that has captured the imagination of the public.

Glossary

absolute velocity: A null quantity because the standard of reference by which this quantity could be measured has not been found. A hypothetical quantity which could be determined by measuring the relative velocity of the aether, or the relative velocity of any alternative medium that could serve as an absolute inertial frame of reference.

aliasing: An illusory wave pattern caused by under-sampling.

amplitude: The height, intensity, or magnitude of a wave. Amplitude manifests as loudness, with respect to sound, and as brightness, with respect to light.

angular momentum: A vector which represents the speed at which an object is spinning along with the direction in which it is spinning.

asymptote: A dotted line which is used as a graphical representation of infinity.

classical object: A macro object. Every object that is not a quantum object (e.g., people, cars, planets, stars). Objects which people can detect with their senses. An object which behaves according to the deterministic laws of classical mechanics but not according to the probabilistic laws of quantum mechanics.

eigenstate: The measured property of a quantum system. An eigenstate has a defined value.

frame-dependent displacement: The total length of the path taken by a vessel as it travels between object A and object B, where we must also consider the distance travelled by object A, and by object B, relative to an arbitrary, inertial frame of reference, during the time that the vessel is in transit. Einstein's concept of the length of the path taken by the light as it bounces between the mirrors of a light clock is an example of a frame-dependent displacement.

fudge factor: An ad hoc variable that is introduced into a mathematical model in order to bring theory into alignment with observation. The duct tape that holds an inadequate theory together.

gravitational mass: The aspect of mass which interacts with gravity. Gravity causes massive objects to orbit around their mutual centre of mass.

gravitational time dilation: Time dilation that is attributable to the force of gravity.

hairy ball theorem: The assertion in algebraic topology that it is impossible to comb all of the hairs on a hairy ball continuously and have them all lay flat. This theorem was advanced by Henri Poincaré.

inertial mass: The aspect of mass which interacts with inertia. Inertia makes it harder to push, pull, or resist the motion of a massive object.

inertial time dilation: Kinematic time dilation. Relativistic time dilation. The slowing of a clock or of the aging process, that is attributable to inertial motion.

jerk: A measure of the rate of change in acceleration over time.

latent light: Light which includes manifest light as well as the rest of the light that cannot be observed by a particular observer because its velocity is outside of the range of values that the observer can interact with, from their location in velocity space. Light which includes all of the light that is and is not selected by an observer.

length contraction: A reduction in the apparent size of an object along the direction of its relative motion, in accordance with the Lorentz transformation equation for length contraction.

light bubble: A hollow sphere in velocity space with a radius of 299,792 km/s. The observer is always located at the centre of their light bubble, in velocity space. The shape that manifest light takes in velocity space.

longitudinal wave: A wave that has a polarization which is parallel to the direction of its motion (e.g., sound waves).

Lorentz transformation equation: An equation that produces a value (*gamma*) which represents the degree of inertial time dilation, length contraction, or inertial mass increase, as a function of a relative velocity.

manifest light: Light which occupies the light bubble of an observer in velocity space. The light which an observer can observe and interact with due to its relative velocity.

moiré pattern: An illusory, wavelike pattern that is created by overlapping transparent gradients.

orbital velocity: The velocity at which one object orbits around another object.

polarization (light): The dimension along which the vibrating motion of a wave takes place.

parallax: A method of estimating the distance to an object through observing the degree of change in its apparent position, relative to the background, that results from viewing it from different locations.

quantum superposition: Wavelike behaviour exhibited by a particle. A position that is loosely defined or uncertain. A probability distribution of potential observations which are unpredictable in principle. Occupancy of multiple, contradictory positions, or properties, at the same time.

quantum object: An object which behaves according to the laws of quantum mechanics (e.g., an elementary particle or an atom). A particle which displays wavelike or indeterminate behavior, such as superposition.

selection mechanism: The process by which a superposition of states reduces to a single state. The mechanism which defines a property by selecting from multiple alternatives.

spacetime: A hybrid manifold which combines three dimensions of space with time, in accordance with the Lorentz transformation equation for time dilation. Manifest light occupies a light cone in spacetime.

superwave: A wave that travels through a medium which occupies a superposition in velocity space.

supervelocity: A superposition in velocity space. The occurrence of moving at multiple speeds and in multiple directions, at the same time.

transverse wave: A wave that has a polarization which is perpendicular to the direction of its motion (e.g., the ripples in a pond).

true distance: A distance between two objects which does not depend on the inertial frame of reference of an observer. Our intuitive sense of what distance or size really means. What a tape measure is designed to measure. The stable distance between the two mirrors in the light clock thought experiment, is an example of a true distance.

vector: A mathematical representation which combines a magnitude with a direction.

velocity: A vector which combines a speed with a direction.

velocity orbit: The path that is traced by an object through velocity space as it orbits another object.

velocity space: A three-dimensional space where all vector distances between points represent relative velocities, all scalar distances between points represent relative speeds, and all points represent velocities. Objects which are not moving relative to one another will share the same location in velocity space, and position is not represented in velocity space.

Bibliography

Born, Max. "The Statistical interpretation of Quantum Mechanics." Nobel Lecture, December 11, 1954.

Cairoli, Klages, and Baule. "Weak Galilean Invariance as a Selection Principle For Coarse-Grained Diffusive Models." Proceedings of the National Academy of Sciences (PNAS), 2018.

Daimon, Masahiko, and Masumura, "Measurement of the Refractive Index of Distilled Water From the Near-Infrared Region to the Ultraviolet Region," Applied Optics (2007)

Albert Einstein, "On the Electrodynamics of Moving Bodies," Annalen der Physik (1905)

Einstein, Podolsky, and Rosen, "Can Quantum Mechanical Description of Physical Reality Be Considered Complete?" Physical Review (1935)

Richard Feynman, *Six Easy Pieces: Essentials of Physics Explained by Its Most Brilliant Teacher.* Basic Books, 1963.

Richard Feynman, *Six Not-So-Easy Pieces: Einstein's Relativity, Symmetry, and Space-Time.* Basic Books, 1963.

Hafele and Keating, "Around-the-World Atomic Clocks: Observed Relativistic Time Gains," Science (1972)

Todd Jaquith, "Pulsar Finally Found in the Andromeda Galaxy," Futurism (2016)

Kane, Guan, and Banks, "The Limits of Human Stereopsis in Space and Time," The Journal of Neuroscience (2014)

Susanna Kohler, "Determining Our Motion Through the Galaxy," American Astronomical Society (2016)

Lima, Nascimento, Cavalcanti, and Ostermann, "Louis de Broglie's Wave-Particle Duality: From Textbooks' Black boxes to a Chain of Reference Presentation," Revista Brasileira de Ensino de Física (2020)

Lincoln, Don. *Is relativistic mass real?* YouTube uploaded by Fermilab, 5 Sept. 2017. https://www.youtube.com/watch?v=LTJauaefTZM

Lincoln, Don. *Why does light slow down in water?* YouTube uploaded by Fermilab, 20 Feb. 2019. https://www.youtube.com/watch?v=CUjt36SD3h8

Lulu Liu, "The Speed and Lifetime of Cosmic Ray Muons," Massachusetts Institute of Technology (2007)

Karl Popper, "Science as Falsification," Conjectures and Refutations (1963)

G. Mather, "The Use of Image Blur as a Depth Cue," Perception (1997)

Daniel McNulty, "The Bayesian Method of Financial Forecasting," Investopedia (2022)

Michelson and Morley, "Relative Motion of the Earth and the Luminiferous Ether," American Journal of Science (1887)

Denis Michel, "Calculation of Distances and Distance-Redshift Relationships Under the Different Modes of Space Expansion," (2015) Hal01221674v2

Gregory Naber, *The Geometry of Minkowski Spacetime: An Introduction to the Mathematics of the Special Theory of Relativity.* Courier Corporation, 2003.

NASA. Dark Energy, Dark Matter https://science.nasa.gov/astrophysics/focus-areas/what-is-dark-energy

Thomas Nagel, *The View from Nowhere.* Oxford University Press, 1986.

R. Oerter, *The Theory of Almost Everything: The Standard Model, the Unsung Triumph of Modern Physics.* Penguin Group, 2006.

Henri Poincaré, "Sur les courbes définies par les équations différentielles," Journal de Mathématiques Pures et Appliquées," (1885)

C.D.C. Reeve, *Plato Republic.* Hackett Publishing Company, Inc., 1992.

Rossi, Bruno, and Hall, "Variation of the Rate of Decay of Mesotrons with Momentum," Physics Review (1941)

Rubin, Vera C., and Ford, "Rotation of the Andromeda Nebula from a Spectroscopic Survey of Emission Regions," Astrophysical Journal (1970)

Ethan Siegel, "How Do Photons Experience Time?" Forbes Magazine (2016)

Ethan Siegel, "Why 21 cm Is the Magic Length for the Universe," Big Think (2022)

Katie Spalding, "The Most Important Number? It's 137. This is Why," IFLScience (2022)

The definition of the meter https://www.nist.gov/si-redefinition/meter

The definition of the second https://www.nist.gov/si-redefinition/second-introduction

Tirole, Romain; Vezzoli, Stefano, et al. "Double-Slit Time Diffraction at Optical Frequencies," Nature Physics (2023)

Ernie Tretkoff, "February 1927: Heisenberg's Uncertainty Principle," American Physical Society News (2008)

Young, Thomas. "The Bakerian Lecture: Experiments and Calculation Relative to Physical Optics," Philosophical Transactions of the Royal Society of London (1804)

Zanella, Giovanni "The Physics of the Longitudinal Light Clock," Università di Padova, Italy